Alcoholism

THE FACTS
Second edition

BY

DONALD W. GOODWIN M.D.

Professor of Psychiatry, The University of Kansas

Oxford New York Tokyo

OXFORD UNIVERSITY PRESS

1994

Oxford University Press, Walton Street, Oxford OX2 6DP
Oxford New York Toronto
Delhi Bombay Calcutta Madras Karachi
Kuala Lumpur Singapore Hong Kong Tokyo
Nairobi Dar es Salaam Cape Town
Melbourne Auckland Madrid
and associated companies in
Berlin Ibadan

Oxford is a trade mark of Oxford University Press

Published in the United States
by Oxford University Press Inc., New York

First edition published 1981
Second edition published 1994

A catalogue record for this book is available from the British Library

Library of Congress Cataloging in Publication Data
Alcoholism / by Donald W. Goodwin.—2nd ed.
(The Facts) (Oxford medical publications)
Includes index.
1. Alcoholism. I. Title. II. Series. III. Series: Facts
(Oxford, England)
RC565.G638 1994 616.86'1—dc20 93-33957
ISBN 0 19 262339 7 (Hbk)
ISBN 0 19 262338 9 (Pbk)

Typeset by Downdell, Oxford
Printed in Great Britain by
Biddles Ltd, Guildford & King's Lynn

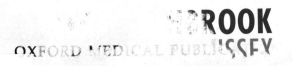

Alcoholism

THE FACTS

ALSO PUBLISHED BY OXFORD UNIVERSITY PRESS

ALSO BY DONALD W. GOODWIN

Is alcoholism hereditary? (second edition)

Anxiety

Psychiatric diagnosis (fourth edition)
co-authored with S.B. Guze

Phobia: the facts

Longitudinal research in alcoholism
co-authored with K.T. VanDusen and S.A. Mednick

Alcoholism and affective disorders
co-edited with Carlton Erickson

Alcohol and the writer

Preface

This book is written for people who worry about drinking, their own or somebody else's. For the millions of people who have alcoholism in the family, it may have particular appeal.

There is a lot of misinformation about alcoholism. For example, many people, including doctors, take an excessively pessimistic view of the problem. They believe alcoholics never get well. This is untrue.

Many have heard that alcohol causes brain damage. This has not been proved. The newspapers are full of stories about an increase of alcoholism among teenagers and women. This is *impossible* to prove, because nobody knows how many were alcoholic in the past.

I hope to correct some of these misconceptions. At the same time I realize that, as more is learned, today's facts may be tomorrow's fictions.

I have spent most of my professional life doing studies about alcohol and alcoholism. In recent years the studies have dealt more and more with the possible relationship of heredity to alcoholism. This led to a book, *Is alcoholism hereditary?* (Random House, 1988) written mainly for scientists. There is no technical language in the present book. Nor are sources of information given, although most of the information *has* sources. These are scientific studies, by and large, with a certain amount of personal opinion, which was inevitable.

This is not, I hope, a preachy book. I am convinced that alcoholism is an illness and not a vice. I do not believe victims of alcoholism got that way because they exercised free will and chose to get that way.

I don't particularly like the word 'alcoholism'. I use it because it's the word most people use. Those unhappy with the word may substitute 'alcohol dependence'. For brevity, I refer to the alcoholic (the alcohol-dependent individual) in the male gender and the spouse in the female gender, except in the chapter on the woman alcoholic. There are many people who, at some time in their life, have had problems from drinking but do not fit the definition of alcoholism in Chapter 5. They are called, for want of a better term, 'problem drinkers'. They, too, are dealt with in this book.

As for my own philosophy about drinking, it is captured rather nicely by the quotation introducing Section 1.

Kansas City   DWG
March 1994

Contents

To Sally

SECTION ONE

Alcohol

You have asked me how I feel about whisky. All right, here is just how I stand on this question:

If, when you say whisky, you mean the devil's brew, the poison scourge, the bloody monster that defiles innocence, yea, literally takes the bread from the mouths of little children; if you mean the evil drink that topples the Christian man and woman from the pinnacles of righteous, gracious living into the bottomless pit of degradation and despair, shame and helplessness and hopelessness, then certainly I am against it with all of my power.

But, if when you say whisky, you mean the oil of conversation, the philosophic wine, the stuff that is consumed when good fellows get together, that puts a song in their hearts and laughter on their lips and the warm glow of contentment in their eyes; if you mean Christmas cheer; if you mean the stimulating drink that puts the spring in the old gentleman's step on a frosty morning; if you mean the drink that enables a man to magnify his joy, and his happiness, and to forget, if only for a little while, life's great tragedies and heartbreaks and sorrows, if you mean that drink, the sale of which pours into our treasuries untold millions of dollars, which are used to provide tender care for our little crippled children, our blind, our deaf, our dumb, our pitiful aged and infirm, to build highways, hospitals and schools, then certainly I am in favour of it.

This is my stand. I will not retreat from it; I will not compromise.

—Address to the legislature
by a Mississippi state senator
in 1958

1 *Alcoholic beverages*

Let us begin at the beginning: with yeast.

When yeast grows in sugar solutions without air, most of the sugar is converted (fermented) into carbon dioxide and alcohol. Carbon dioxide makes the solution bubble ('fermentation' comes from the Latin word for 'boil') and makes champagne corks pop. The alcohol is excreted. Most drinkers do not know they are drinking yeast excrement. Would it matter?

It matters for yeast. When the alcohol concentration reaches about 12 or 13 per cent, the yeast dies of acute alcohol intoxication. This is why unfortified wines, produced by fermentation alone, have alcohol concentrations of no more than 12 or 13 per cent. Sherry, port, and other fortified wines have alcohol added.

As a rule, people do not drink just alcohol. They drink alcoholic beverages. Alcoholic beverages are mostly water and a two-carbon alcohol molecule called ethyl alcohol or ethanol. Tiny amounts of other chemicals, called congeners, also are present. They provide most of the taste and smell and all of the colour, if any.

Because of congeners, beer can be distinguished from brandy, although both consist almost entirely of ethyl alcohol and water. Congeners, depending on the beverage, include varying amounts of amino acids; minerals; vitamins; a one-carbon alcohol called methanol or 'wood alcohol'; plus the 'higher' alcohols with more than two carbons, otherwise known as fusel oil.

Even in small quantities, wood alcohol and fusel oil are poisons. So is ethyl alcohol, but a lot more of it is required to do damage. Is there enough wood alcohol and fusel oil in a cocktail to hurt anyone? Probably not, but no one is sure. To be on the safe side, some people avoid drinks with large amounts of congeners—whisky and brandies—and drink relatively congener-free vodka. They are not aware that vodka often contains more wood alcohol—the notorious blinder—than other beverages.

Although there is almost certainly not enough to blind, wood alcohol certainly does not improve vision.

Congeners vary not only from beverage to beverage but also from brand to brand, and even from bottle to bottle of the same brand. Not all brands (or bottles) of vodka contain relatively high amounts of wood alcohol, but some do. Russian vodka, the favourite of many vodka connoisseurs, sometimes contains considerable amounts of wood alcohol, although, again, not enough to be harmful.

For many years there was a movement in the USA to label alcoholic beverages as 'dangerous to your health'. Indeed they are, if taken in excess. With much opposition from the alcoholic beverage industry, Congress finally passed a law requiring a warning on alcoholic beverages that 'according to the surgeon general', women should not drink during pregnancy, nor should anyone drink who drives a car or operates machinery, followed by a warning that drinking may cause health problems. The warnings are in tiny print and often placed on bottles where they are unlikely to be seen, such as the neck. A more useful label for alcoholic beverages might list the congeners content in the same way that ingredients are listed on food packages. In the small amounts present in beverages, congeners may not be dangerous but heavy drinkers ingest rather a large amount of congeners, and some believe, with uncertain evidence, that some of them contribute to hangover and medical complications from heavy drinking.

Apart from man's contribution—the brewer's art, the cosseted grape—beverages differ according to the sugar source. From grapes, wine; from grain and hops, beer; from grain and corn, whisky; from sugar cane, rum; and originally from the lowly potato, but now mainly from grain, vodka.

Man's great achievement in improving upon yeast's modest productivity was distillation, discovered about AD 800 in Arabia ('alcohol' comes from the Arabic *alkuhl*, meaning essence). Distillation boils away alcohol from its sugar bath and re-collects it as virtually pure alcohol. Then, because pure alcohol is pure torture to drink, water is added back, so that instead of having

100 per cent alcohol, you have 50 per cent or 100 proof alcohol (proof being one-half per cent).

The alcohol content of most distilled alcoholic beverages is expressed in degrees of proof. This term probably developed from the seventeenth-century English custom of estimating content by moistening gunpowder with the beverage and applying a flame to the mixture. The lowest alcohol concentration that would allow ignition—a concentration of about 57 per cent alcohol by volume—was considered to be 'proof spirits'. British and Canadian regulations are still based on this yardstick; a concentration of 57.35 per cent alcohol is considered to be 'proof spirits', while other concentrations are described as 'over' or 'under' proof.

In many ways alcohol resembles water. In the body alcohol behaves like water. It travels everywhere water travels. Because of its water-like properties, ethyl alcohol can be accommodated by the body in vastly greater amounts than any other drug. A person's blood can consist of one half of one per cent of alcohol without producing death or even unconsciousness.

2   Alcohol in the body

What happens to alcohol when you drink it? Essentially the same thing that happens if you don't drink it. It turns to vinegar.

When alcohol 'sours' in the open air, bacteria are responsible.* To become vinegar (acetic acid) in the body, alcohol needs two enzymes: alcohol dehydrogenase and aldehyde dehydrogenase. The first is located in the liver in surprisingly large supply. Surprising because, as far as we know, alcohol dehydrogenase does nothing except metabolize alcohol. It is there in all mammalian livers—in the horse's in particular plentitude. Why? Did God anticipate that someday a mammal like man would develop a taste for alcohol and need a way to dispose of it? Or did it happen that millennia ago horses and other vegetarians ate fermenting fruit lying on the ground and their obliging livers evolved a helpful enzyme?†

Nobody knows, but it is a nice enzyme to have in any case. It disposes of 100-proof distilled spirits at the rate of about one ounce per hour, slow enough to soak the brain without, one hopes, pickling it.

Alcohol dehydrogenase is located mainly in the liver, but not exclusively. Small amounts are also found in the lining of the stomach. Men have more alcohol dehydrogenase in their stomachs than woman. It has long been noted that women seem to become more intoxicated than men on similar amounts of alcohol. The reason may be they lack sufficient alcohol dehydrogenase—the enzyme that breaks down alcohol—in their stomachs. With men, the alcohol is partly broken down in the

---

* The reason that fortified wines and distilled spirits do not 'sour' in the open air is that they have concentrations of alcohol above 12 or 13 per cent, which is as lethal for bacteria as it is for yeast.

† A minute amount of ethyl alcohol is produced in the gastrointestinal tract by bacteria, and perhaps this accounts for alcohol dehydrogenase in the liver. Infinitesimal amounts of alcohol are produced by normal metabolic processes in the body. If these sources are the reason that the alcohol enzyme is present in such large quantities, it is clearly a case of biological overkill.

stomach and thus less passes into the small intestine from which diffusion into the blood stream is rapid. Women deliver more straight alcohol to the small intestine than men, resulting in higher blood levels on the same amount. This may partly account for differences between men and women in their responses to alcohol, the subject of Chapter 8.

Between alcohol and acetic acid is an intermediate step, which is why a second enzyme is required. The intermediate chemical is an aldehyde and very toxic. Again, nature saves the day. The enzyme that destroys the aldehyde is found not just in the liver but throughout the body. It quickly turns the aldehyde into harmless acetic acid. Fed into the body's normal metabolic machinery, acetic acid becomes carbon dioxide and water, burning or storing seven calories per gram of alcohol in the process.

Vinegar is harmless, but the process that produces it may not be. In being oxidized, alcohol is progressively stripped of hydrogen atoms, which must go somewhere. Where they go results in some interesting biochemical changes which may or may not be harmless (the evidence is not yet clear). Some of the changes are:

1. There is an increase in lactic acid. This is interesting because a connection between increased lactic acid and anxiety attacks has been observed, and heavy drinking is also associated with anxiety attacks.

2. There is an increase in uric acid. This is interesting because increased uric acid is associated with gout, and gout for centuries has associated with alcohol.

3. There is an increase in fat—not the slow increase that comes from calories (those seven calories per gram) but a rapid increase from the oxidation of alcohol. This fat is seen mainly in the liver or blood. One night of serious drinking (say, six or seven whiskies) discernibly increases the fat content of the liver. The liver will be fattier still if fatty food is also eaten.

Is a fatty liver bad? Admittedly it does not sound good, but on the other hand, the connection between a fatty liver and liver diseases, such as hepatitis and cirrhosis, is unclear. For one thing, the fat goes away soon after the drinker stops drinking. Also, most people drink but most do not develop liver disease. Among those very heavy drinkers we call alcoholics, perhaps only 5 or 10 per cent develop liver disease, although presumably their livers are fatty most of the time. On the other hand, people who develop a particular type of liver disease called Laennec's cirrhosis usually *are* heavy drinkers.

Many disorders associated with heavy drinking are apparently caused by malnutrition, but this may not be true of cirrhosis. Laennec's cirrhosis has been produced in well-nourished baboons after four years of drunkenness. Most of the drunk baboons, however, only had fatty livers, and controversy still thrives about whether alcohol alone causes cirrhosis. Obviously it does not in every case.

Intoxicating amounts of alcohol also increase fat in the bloodstream. In high enough doses, particularly combined with a fatty meal, alcohol may produce *visible* fat in the blood. The plasma takes on a faint milky tinge, probably a more frightening development for most people than the notion of having fat in their liver. Still, while it does not sound good, nobody knows how bad it is, and possibly it is not harmful at all.

For years people have looked for ways to speed up the elimination of alcohol from the body. The theory was that if alcohol went away faster, some of the ominous things described above might not occur. They probably would occur nevertheless, but speeding up alcohol's elimination would serve one purpose: it would shorten the period of intoxication.

Many things have been tried—insulin, caffeine, exercise—but only one has worked. Fruit sugar (fructose) in large doses definitely speeds the elimination of alcohol. Unfortunately the dose required is so large it is sickening, and most people prefer to remain drunk.

To some extent, alcohol intoxication can be reduced *without* lowering blood alcohol. Drugs called narcotic antagonists (antagonists because they block the effects of morphine and other narcotics) also seem to prevent impairment from alcohol. The antagonist has to be given by intravenous injection, which poses a practical problem for bar patrons preparing themselves for a safe drive home. Nevertheless, this finding has interesting implications for two effects of alcohol which have never been explained.

These effects are *analgesia* (relief of pain) and *euphoria*. Here, first, is a theory about analgesia.

The brain contains substances resembling morphine. There is speculation that the body normally responds to painful stimuli by increasing the activity of these natural morphine-like substances. Different people react differently to pain and the reason may be that some of us are 'born' with more 'morphine' in our brain than others. What does this have to do with alcohol? One line of thought holds that alcohol produces analgesia—and possibly euphoria—by stimulating the release of these normally occurring morphine-like substances.

Alcohol also may have morphine-like effects (analgesia and euphoria) in another way. A breakdown product of alcohol combines with a normally occurring chemical in the brain to produce a substance resembling morphine. Thus, alcohol may cause analgesia and intoxication in two ways. It may stimulate the release of morphine-like substances or it may produce new morphine-like substances in the brain. Both may happen, or neither. Alcohol and morphine states are quite different, as are the withdrawal syndromes.

3 Alcohol and behaviour

The effects of any drug depend on the dose. The chance of death occurring from a sip of beer is remote. A quart of whisky drunk in an hour will kill most men. This dose–effect rule applies to any substance a person consumes. Everything is either a poison or harmless depending on the dose. People die from drinking too little water, and from drinking too much. A little strychnine may even be good for you (it helps rats concentrate). People who still argue about whether marijuana is more or less harmful than alcohol, should ask the following: Does alcohol refer to a bottle of low-alcohol beer or to a quart of whisky? Does marijuana refer to Kansas hemp (the cannabis equivalent of low-alcohol beer) or to Moroccan hashish (the whisky counterpart)? Drug comparisons are senseless except in terms of amount. (Based on existing evidence, and again taking dose into consideration, marijuana may have about the same addictive potential as alcohol. One in ten marijuana users become dependent on the drug to the point where it interferes with their lives. About one in ten drinkers become dependent on alcohol.)

Quantity is not everything. To the amount consumed must be added other factors in determining a drug's effects:

1. *Concentration of alcohol in the blood*

What really counts is not how much alcohol a person drinks but how much gets into the bloodstream. This in turn depends on many things.

As mentioned, some alcohol is absorbed through the stomach wall but most reaches the bloodstream through the small intestine. Between the stomach and small intestine is a muscular ring called the pyloric valve. (Anatomists point out that the pyloric valve is not a true valve, but merely functions as a valve. This is like saying that the works of Shakespeare were not written by Shakespeare but by another man of the same name.) When the valve clamps shut, as may happen when jolted by a straight shot of whisky, the alcohol remains in the stomach, where it is absorbed at a very slow pace. For people with sensitive pyloric valves, a strong shot of whisky or an extra dry martini may be self-defeating if a quick effect is the goal.

For rapid absorption, it is important that the alcohol reach the small intestine in the highest possible concentration and the shortest possible time. People who have had their pyloric valves removed surgically, as in the treatment of ulcers, find they get drunk faster than previously. Without a pyloric valve to slow down the passage of alcohol, it swooshes into the small intestine and thence into the bloodstream.

Other factors influencing absorption include the presence or absence of food in the stomach and the type of beverage. When alcohol has to compete with other, larger, and often more aggressive molecules in crossing the gastric and small intestinal membranes, it often does not fare well. Figure 3.1 illustrates this. Alcohol in the form of different beverages was given to subjects with and without food. Although the same amount of alcohol was consumed over the same length of time, the blood-alcohol concentration varied greatly. Gin on an empty stomach produced a peak blood-alcohol level above the legal blood-alcohol concentration for drivers in America and most other Western countries. Beer combined with a meal resulted in a peak blood level legally compatible with driving anywhere.

There has been speculation about the difference in alcoholism rates between two wine-drinking countries, France and Italy. Italy, with the lower rate, has a national tradition of drinking wine mainly with meals, while the French tend to drink wine between meals as well as with them. When wine and spaghetti compete for transport across the intestinal wall, it is not surprising that spaghetti, finishing the race first, will prevent the wine from making much headway.

Mixing food and alcohol produces a slight increase in the oxidation of alcohol and hence its removal from the blood-stream. This may partly explain the lower blood-alcohol level when food and alcohol are combined. It also may partly explain the known reluctance of alcoholics to eat while drinking, since

presumably the alcoholic has no strong desire to remove alcohol
from his bloodstream any faster than necessary.

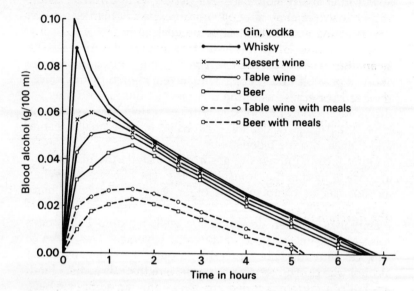

Fig. 3.1 Typical blood-alcohol curves resulting from ingestion of
various spirits, wines, and beer, each at amounts equivalent to 0.6 g
of alcohol per kilogram of body weight (3 oz of vodka, three beers, or a
large glass of wine). Redrawn from *Alcoholic Beverages in Clinical Medicine*
by C. D. Leake and M. Silverman. Copyright © 1966 by Year Book
Medical Publishers, Inc., Chicago. Used by permission.

2. *Rate of absorption*

In addition to how much alcohol is in the blood, it matters how
quickly it got there. In general, the faster the rate of absorption,
the more striking the effect. This may help explain the popularity of a dry martini before meals.

3. *Duration of drinking*

The body adapts rapidly to chemical insults, including alcohol.
The longer there is alcohol in the blood, the more its effects

diminish. In practical terms (for a secretary or writer, anyway), if you make five errors per minute while typing sober, you may make 15 errors per minute while typing with a certain blood-alcohol concentration after one hour of drinking, but only seven errors at the same concentration after five hours of drinking. After five hours you may not care how many you make, but that is another consideration.

4. *The slope effect*

Any drinker can tell you he feels better getting drunk than he does sobering up. That is, as the blood-alcohol level climbs from A to B to C, he may feel euphoric at B and C, but as the blood level falls from C to B to A, not only is there no euphoria at B, there is actual discomfort, presaging the hangover to come at A. This 'slope' effect is closely related to and hard to separate from the duration effect.

5. *Tolerance*

As people drink more over days, months, and years, they gradually *need* to drink more to obtain the same effect. This is called tolerance. Its importance is often exaggerated. A seasoned alcoholic at the prime of his drinking capacity may be able to drink, at most, twice as much as a teetotaller of similar age and health. Compared with tolerance for morphine, which can be manyfold, tolerance for alcohol is modest. (Apparently, alcohol in the same amount is as lethal for heavy drinkers as for light drinkers, although the experimental studies required to prove this have not been done for obvious reasons. Tolerance to opiates—including morphine and heroin—includes tolerance to the lethal level.)

More striking than 'acquired' tolerance may be inborn tolerance. Individuals vary widely in the amount of alcohol they can tolerate independent of drinking experience. Some people, however hard they try, cannot drink more than a small amount of alcohol without developing a headache, upset stomach, or dizziness. They rarely become alcoholic but deserve no credit for it. Their 'alcohol problem' is that they *cannot* drink very much.

Others seem able to drink large amounts with hardly any bad effects. It appears they were born with this capacity and did not develop it entirely from practice. They *can* become alcoholic and some do.

Differences in tolerance for alcohol apply not only to individuals but to racial groups. For example, many Orientals develop flushing of the skin, sometimes with nausea, after drinking only a little alcohol. For obvious reasons, alcoholism is rare in these groups. American Indians are also said to be intolerant of alcohol but the nature of the intolerance is more ambiguous and does not appear to discourage heavy drinking.

6. *Set and setting*

Any drug response that involves thinking and mood is bound to be influenced by expectation. Alcohol is no exception. If a person believes alcohol will improve his mood, diminish fatigue, make him feel sexy, or have other salutary effects, the chances that these pleasant changes will occur may be improved. The same goes for expectations of unpleasant changes.

In medicine this is called the placebo effect: drugs tend to do for people what they expect them to do. It presumably has little or no role in the treatment of pneumonia with antibiotics—although one cannot be certain—but it has powerful implications for treating psychiatric disorders. Sometimes sugar pills help almost as much as expensive and potentially toxic tranquillizers or antidepressants.

It is the gap between the sugar pill's performance and that of the 'active' drug that justifies the prescription of the drug, and sometimes the gap is quite small. Placebos can even produce side-effects—headaches, nausea, a rash. Everyone is a little susceptible to suggestion, some more than others.

It is difficult to know how much the effect of alcohol in any given person on any particular occasion is influenced by expectation, or what psychologists call set. Presumably, the stronger the dose, the smaller the placebo effect. But it is a common laboratory and social observation that some people get 'drunk' on very little alcohol. (Richard Nixon, according to

Henry Kissinger, was such a person.) This may be because they want or expect to get drunk quickly.

But set refers to more than expectation. If a person is tired, alcohol may have more of an anti-fatigue effect than usual. If he is hungry, it may make him more hungry (or less). If his mood is good, it may become better. If bad, worse. All of this refers to set—the psychological and physical state of the person at the time he proceeds to drink.

Set, to a considerable extent, is linked to setting. Where is the person drinking? With whom? If he enjoys the people he is with, he may also enjoy the alcohol more. If the occasion is a celebration, a drink may have a livelier effect than would the same amount taken routinely before dinner.

Alcohol is said to make people talk louder, and this often seems true. On the other hand, two men on a deer hunt, taking a nip of scotch to warm up, may talk more softly than usual.

The importance of set and setting in shaping a person's response to alcohol should not be underestimated, although it is difficult to study their relative influence at a given time.

The four stages of intoxication

There is an old saying that alcohol affects a person in four ways. First, he becomes jocose, then bellicose, then lachrymose, and finally comatose.

Comatose he does indeed become if he drinks enough, but the other three stages are not inevitable. Some people hardly feel jocose at all. One reason may be that they do not want to feel jocose. Their reasons for drinking may be purely social: others drink, so do they.

Many people become argumentative when they drink and some combative, but these responses are strongly influenced by social circumstances. The legendary bar-room brawl usually occurs in lower-class bars and is a rarity in upper-class saloons. Countless parties are held nightly in middle-class suburbia and, although drinking is common, fighting is not.

This is not to deny that drinking may bring out the beast in man. Alcohol is involved in at least half of all homicides in the USA, with either the attacker, the victim, or both under the influence. This probably explains why more murders occur on Saturday night than on any other evening (the fewest occur on Tuesdays). Again, the connection between alcohol and belli-cosity has class overtones, since most murders occur in the lower and lower-middle classes, including drug dealers who kill other drug dealers. It has been suggested that one reason fights occur in bar-rooms is that rarely are so many people thrust together so closely for such long periods, with hardly anything to do but talk, drink, and fight.

One of the paradoxes about alcohol is that people sometimes cry when they drink. Why, then, drink? Isn't the whole point of drinking to feel happier? The fact is, even though some people become anxious and depressed when they drink, they do not give up drinking, which challenges the widely held assumption that people drink mainly to feel less anxious and depressed. The motives for drinking, in truth, are complex and inscrutable, with no single explanation sufficing for all circumstances.

Alcohol is often described as a 'depressant' drug that depresses first the 'higher' centres in the brain and then down-wardly anaesthetizes the brain until finally, in lethal dosage, it snuffs out life itself by depressing the respiratory centre at the base of the brain. This, like most things said about alcohol, is an oversimplification.

What is alcohol 'depressing'? Usually not activity. Most people get a 'lift' from alcohol and many become more animated and active. Nerve fibres 'fire' about as readily in an alcohol solution as they do otherwise, unless the concentration is far above what most people can achieve by drinking. It is some-times said that by depressing the 'higher' centres of the brain, alcohol releases the 'lower' centres and that this is why people are more uninhibited when they drink—the 'animal instincts' are released. The problem is that studies do not support the theory of the top-to-bottom action of alcohol. Co-ordination, a 'lower' function, often is impaired at lower doses of alcohol than is memory, a 'higher' function.

Again, dosage is crucial. Alcohol in rather small dosage may improve certain types of performance. Apparently this is most likely to occur in activities where the person is not very proficient and where the effects of increased confidence might be expected to show up. If he does poorly at hitting the target on a firing range, he may improve somewhat after several drinks of alcohol. On the other hand, if he does well normally, his performance may fall off when he takes in small amounts of alcohol. Nevertheless, in moderate to high amounts, alcohol usually diminishes function across the board.

An interesting exception to this general rule has emerged in several studies. Apparently if a person learns certain things, such as word lists, while intoxicated—even severely intoxicated—he will remember them better when reintoxicated than when sober. Called 'state-dependent learning', this is one of the few exceptions to the overall impairing effect of alcohol at moderate and high dosages.

Alcohol does something else that is almost unique among drugs. It produces a classical amnesia called 'blackout'. While drinking, the drinker does highly memorable things but cannot remember them the next day. Many social drinkers have had this experience, but it occurs most frequently in alcoholics.

Some misconceptions

Some of the physical effects of alcohol should also be mentioned, if only because there are misconceptions about them.

1. It is generally known that alcohol increases urination. It is generally not known that the increase is temporary, and that after a fairly short period of drinking the need to urinate decreases. On the morning after a night of heavy drinking, a person may not urinate at all. No explanation is available.

2. It is commonly believed that alcohol causes dehydration. It does not. When a person has a dry mouth and thirst after an evening of drinking, it may be because of the astringent effect of alcohol on the mucous membranes of the mouth. If anything,

heavy drinkers may be overhydrated because of the large volume of fluid they consume.

3. It is generally known that alcohol produces a feeling of bodily warmth and therefore is just the thing for Saint Bernards to carry around their necks in barrels and for old boys to have at a frosty football game. Alcohol produces a feeling of warmth because it dilates blood vessels in the skin, which is why drinkers have red noses. However, the warmth can be harmfully illusory. A person's resistance to the effects of severe cold, such as frostbite, is in no way increased by alcohol, although the victim may temporarily think it is.

4. It has been reported that alcohol causes cancer. The story seems to be this: if a person is a heavy smoker and also a heavy drinker, he is more likely to develop cancer of the throat, larynx, and oesophagus than if he is simply a heavy smoker. There is little evidence that alcohol *alone* causes cancer. Its role in promoting cancer of the head and neck is a matter of speculation. Alcohol may increase the solubility of carcinogens from tobacco smoke. A second possibility is that alcohol in large amounts depresses the body's immune response (which it does) and thereby lowers the body's resistance to cancer agents as well as infections.

One argument against the latter theory is that alcoholics do not seem especially likely to develop cancer of other organs in the body (except possibly cancer of the liver). Beer drinkers may have a higher incidence of cancer of the rectum and colon than non-beer drinkers. If true, it may not occur because of the alcohol in beer, which contains many substances besides alcohol. One of these may be the culprit. Another argument against the lowered-resistance theory comes from the frequent observation by recovered alcoholics that, 'When I was drinking I never had a cold. Now I have them all the time.' This may be so, but the real truth may be that minor ailments are ignored in the haze of intoxication, and it should be said in defence of the lowered-resistance theory that heavy drinking went hand-in-hand with tuberculosis back when tuberculosis was common, just as TB and AIDS are now closely associated.

5. It is said that some people are 'allergic' to alcohol. Members of Alcoholics Anonymous sometimes use the word 'allergy' in a metaphorical sense to explain their addiction. Until recently nobody thought alcohol was an allergen in the usual sense. Allergens are usually proteins, while alcohol is, well, alcohol. New evidence, however, suggests that alcohol in some people may indeed cause an allergic response. This refers to the tendency of Orientals, previously mentioned, to develop a blotchy red rash, particularly on the head, neck, and trunk, after drinking a small amount of alcohol. My colleagues and I found that antihistamines prevent the alcohol-induced flush from occurring. Antihistamines of course are famous for their therapeutic powers in such allergies as hayfever, but this does not prove that the Oriental flushing response is *like* hayfever. Its cause remains (if you will forgive the expression) inscrutable.

4 *Alcohol through the ages*

Six millenniums ago at the Sumerian trading post of Godin Tepe in what is now western Iran people were drinking beer and wine. In 1992, chemists analysed a residue in pottery jars found in the ruins of Godin Tepe and identified the stuff as wine or beer. They were found in the same room. A scientist said, 'I think a lot of serious drinking was going on there.' The Sumerians were among the first people to develop a complex, literate society of prospering city-states based on irrigation, agriculture, and widespread trade.

Beer-making began almost as soon as (or even before) people domesticated barley to make bread in the early transition by the Mesopotamians to agriculture around 8000 BC. A longstanding debate in archaeology centres on the question of which came first after the domestication of barley: beer or bread?

There is other evidence that alcohol goes back at least to palaeolithic times. This derives from etymology as well as from studies of Stone Age cultures that survived into the twentieth century.

Available to palaeolithic man, presumably, were fermented fruit juice (wine), fermented grain (beer), and fermented honey (mead). Etymological evidence suggests that mead may have been the earliest beverage of choice. The word *mead* derives—by way of *mede* (Middle English) and *meodu* (Anglo-Saxon)—from ancient words of Indo-European stock, such as *methy* (Greek) and *madhu* (Sanskrit). In Sanskrit and Greek, the term means both 'honey' and 'intoxicating drink'. The association of honey, rather than grain or fruit, with intoxication may indicate its greater antiquity as a source of alcohol.

All but three of the numerous Stone Age cultures that survived into modern times have been familiar with alcohol. 'The three exceptions,' Berton Roueché writes, 'are the environmentally underprivileged polar peoples, the intellectually stunted Australian aborigines, and the comparably lacklustre primitives of

Tierra del Fuego.' Early European explorers of Africa and the New World frequently discovered that alcohol was important in the local cultures. The Indians of eastern North America, for instance, were using alcohol in the form of fermented birch and sugar maple sap.

Alcohol has been used medicinally and in religious ceremonies for thousands of years, but it has a long history of recreational use. Noah, according to the Old Testament, 'drank of the wine and was drunken'. Mesopotamian civilization provided one of the earliest clinical descriptions of intoxication and one of the first hangover cures. Mesopotamian physicians advised as follows: 'If a man has taken strong wine, his head is affected and he forgets his words and his speech becomes confused, his mind wanders and his eyes have a set expression; to cure him, take licorice, beans, oleander . . . to be compounded with oil and wine before the approach of the goddess Gula (or sunset), and in the morning before sunrise and before anyone has kissed him, let him take it, and he will recover.'

One of the few surviving relics of the Seventeenth Egyptian Dynasty, which roughly coincided with the reign of Hammurabi, is a hieroglyphic outburst of a female courtier. 'Give me eighteen bowls of wine!' she exclaims for posterity. 'Behold, I love drunkenness!' So did other Egyptians of that era. Drunkenness was not rare, historians write, and seems to have occurred in all layers of society from the farmers to the gods (or ruling class). Banquets frequently ended with the guests, men and women, being sick, and this did not in any way seem shocking.

Not only descriptions of drunkenness are found in the historical record, but also pleas for moderation. Dynastic Egypt apparently invented the first temperance tract. Moderation was recommended by no less an authority on moderation than Genghis Khan: 'A soldier must not get drunk oftener than once a week. It would, of course, be better if he did not get drunk at all, but one should not expect the impossible.' The Old Testament condemns drunkenness, but not alcohol. 'Give strong drink unto him that is ready to perish,' the Book of Proverbs proclaims, 'and wine unto those that be of heavy hearts. Let him drink, forget his poverty, and remember his misery no more.'

The 'strong drink' of the Bible was probably undiluted wine. 'She hath mingled her wine,' reports Proverbs; a mixture of wine and water was the usual Jewish drink.

Alexander the Great was one of a lung line of heavy drinking generals. According to Plutarch, Alexander was under the influence of alcohol at the time of the burning of the royal palace at Persepolis in 330 BC, seven years before his death. With torch in hand, a drunken Alexander led revellers in a procession in honour of Dionysus and threw the first firebrand, an act he bitterly regretted when sober. The final year of Alexander's life was punctuated with drunken binges and his death may have been hastened by alcohol withdrawal.

Plutarch describes a mass orgy in 325 BC involving Alexander and his Macedonian army: 'Not a single helmet, shield or spear was to be seen, but along the whole line of march the soldiers kept dipping their cups, drinking-horns or earthenware goblets into huge casks and mixing bowls and toasting one another, some drinking as they marched, others sprawled by the wayside.' The history of war is filled with similar scenes.

Alcohol has been the 'intoxicant of choice' in Judaeo-Christian culture. 'To drink is a Christian diversion/Unknown to the Turk and the Persian,' wrote Congreve 300 years ago. It was not *totally* unknown to the Turk and the Persian, but it is true they favoured other intoxicants, notably, the products of the poppy and hemp plant.

One of the myths of our times is that the 'stresses' of modern living have produced a society unusually reliant on alcohol. This is not true. Per capita consumption in the USA was highest, at an estimated six or seven gallons per person, in the early 1800s when whisky and cider were the favourite beverages. One reason was that whisky is more portable than grain, and cider more portable than apples. Portability was important before trains came along, especially if you were Westward bound in a covered wagon.

In the UK, drinking and drunkenness reached a peak during the 'gin epidemic' of the mid-eighteenth century, when gin sold for a few pennies a pint. Probably in no period of history have so many inebriates crowded the streets of a city as occurred in Hogarth's London. Less beer is consumed per capita in the UK now than 100 years ago. Consumption of wine *was* decreasing but started upward again in the late 1960s when the import fees were reduced and the European Common Market influence took hold.

Fluctuations in consumption also have been influenced by availability of potable water, the introduction of coffee, tea, and cocoa at prices the population could afford, and the waxing and waning of temperance movements.

Alcohol consumption has gone down steadily in recent years in the USA and most Western countries. There has been no decline, apparently, in Russia and Poland. Today, in the United States, the reported consumption of alcohol (pure alcohol) is a little less than three gallons per person over 14 per annum. This figure is based on tax data. Untaxed sales, such as those on military installations, are not included, so per capita consumption may be underestimated. Also, because consumption estimates are based on the resident population, when residents of one state cross into another state to purchase lower priced alcoholic beverages, the result is a higher per capita consumption figure for the state in which sales occur. Washington, DC, and states with high rates of tourism and business travel have higher reported consumption rates because sales to transients are calculated as consumption by the resident population.

The recent reduction of consumption in the USA has occurred despite the fact that people are drinking less untaxed alcohol. There is less available. Moonshine used to be a booming industry in some back country areas, but no longer is today. Making moonshine alcohol mainly was a small family business, and small family businesses in the USA have declined.

There has been an interesting change in beverage preference. In diet-conscious America people drink more 'light' beer (beer with fewer calories) than regular beer. They also drink more 'white' spirits, such as vodka or gin, than 'brown' spirits (whisky). The explanation for the latter is not clear, unless people have the notion that white spirits are healthier than brown, which is probably not true.

International comparisons are difficult at best. However, it appears that 'wine countries' such as France and Italy consume more alcohol than do countries where distilled spirits are favoured. Israel has the lowest per capita consumption. Ireland, contrary to its popular image, has a lower consumption rate than the UK. The USA ranks in the middle with regard to alcohol consumption and Russia (if you can believe the figures) is in the lower third. (You cannot believe the figures. Russians continue to be among the heaviest drinkers in the world, perhaps *the* heaviest.)

Most adults in the USA are light drinkers. About 35 per cent abstain, 55 per cent drink less than three drinks per week, and only 11 per cent consume an average of one ounce or more of alcohol per day.

Drinking patterns vary by age and sex. For both men and women, the prevalence of drinking is highest and abstention is lowest in the 21–34 year age range. Four to five times more males are 'heavy' drinkers compared with females at all ages. For ages 65 years and older, abstainers exceed drinkers in both sexes and only 7 per cent of men and 2 per cent of women are considered heavy drinkers.

The level of consumption varies markedly in different segments of the population. In the USA consumption is greatest in the north-east, lowest in the south. Young white males drink more than any other group in the USA.

The proportion of adolescents who report drinking increases steadily with age, reaching 80 to 90 per cent among the oldest schoolchildren. By that time as many girls report having 'ever' drunk as do boys.

Most alcohol is consumed by a small percentage of people: 70 per cent of the drinking population consume only 20 per cent of the total alcohol consumed; 30 per cent of drinkers consume 80 per cent of the alcohol; and 10 per cent consume 50 per cent.

Consumption must be distinguished from alcoholism. Is the latter increasing?

There is evidence it may be *decreasing*. The distinguished American physician, Benjamin Rush, in 1795 estimated that 4000 Americans died each year from 'over-indulgence in ardent spirits'. Since the population of the country was about four million, this gives a rate of 100 per 100 000. The officially recorded rate of death from alcoholism in the USA today is two per 100 000. Granting that Dr Rush's estimate is suspect, alcoholism in the United States may be less prevalent today than it was 200 years ago.

It's hard to say. There are several problems in estimating the prevalence of alcoholism. One is that few agree on the definition of alcoholism. Also, when a household survey is done, the alcoholics, more than most people, are not home. Neighbourhood bars are rarely included in household surveys.

A third reason to be sceptical about prevalent estimates is that axe-grinding is a potent factor in the production of statistics. Government officials spend much of their time trying to wring money out of reluctant legislators and often are caught in the dilemma of, on the one hand, wanting the prevalence of alcoholism to be low to show they are doing a good job and, on the other hand, wanting it to be high to inspire lawmakers to spend more money on the problem.

In the late 1960s, for example, the US Government announced that there were five million American alcoholics. Now it is said that there are fourteen million. This increase coincides with increased efforts by the Government to study and treat alcoholism. The connection no doubt is coincidental, but since the figures are fictitious anyway it does not matter. No one doubts that alcoholism is common.

Normal drinking

Throughout the ages (and throughout this book), a distinction has been made between normal and abnormal drinking. Before dealing at length with abnormal drinking (drinking that produces problems) something should be said about normal drinking.

How much can you drink and still be 'normal'?

Normal can be defined in several ways. It can be defined as drinking no more than 'society' deems safe and prudent, i.e. normal. Since societies vary in this regard, the definition is not very helpful.

According to another definition, normal drinking is drinking less than is required to produce medical, social, or psychological problems. The problem definition also has problems, as will be discussed in the next chapter.

Finally, attempts are made from time to time to separate normal from abnormal drinking in terms of quantity of alcohol consumed. A nineteenth-century British physician named Dr Anstie proclaimed that normal drinking consisted of drinking no more than three ounces of whisky or half a bottle of table wine or two pints of beer a day (known for years as 'Dr Anstie's limits').

In 1990 the US Government defined normal drinking somewhat more conservatively. It defined moderate drinking (a synonym for normal) as no more than one drink a day for most women and no more than two drinks a day for most men. A 'drink' was defined as twelve ounces of beer, five ounces of wine, or 1.5 ounces of 80-proof distilled spirits. Each of these drinks contain about the same amount of absolute alcohol—1 ounce or 12 grams.

The government points out that even this amount of alcohol should be avoided by some people. These include women who are pregnant or trying to conceive; people who drive or engage in activities that require attention or skill; people taking medications, including over-the-counter medications; recovering alcoholics; and persons under the age of 21.

As noted, women become more intoxicated than men on an equivalent dose of alcohol because, apparently, of a difference of activity of an enzyme in stomach tissues of males and females that breaks down alcohol before it reaches the bloodstream. The enzyme is four times more active in non-alcoholic men than in women. Also, women have proportionally more fat and less body water than men. Because alcohol is more soluble in water than fat, a given dose becomes more highly concentrated in a female's body than in a male's. (Incidentally, body fat increases with age. Some recommend a limit of one drink per day for the elderly.)

Authorities concede that a small amount of alcohol may be beneficial. Alcohol reduces tension, anxiety, and self-consciousness. It promotes conviviality. In the elderly, moderate drinking stimulates appetite, promotes regular bowel function, and improves mood.

In addition, there is considerable evidence that moderate drinking decreases the risk of death from coronary artery disease. In one study, American men who drank three pints per day were less likely to die than men who reported abstinence; they had fewer heart attacks. In another study, one drink a day decreased the risk of coronary heart disease in middle-aged women. In postmenopausal women, the apparent protective effect of alcohol may be explained partly by an alcohol-induced increase in oestrogen levels.

Authorities disagree about whether to recommend that people drink to prevent heart attacks. Even moderate drinking has risks that might offset the benefits. Low levels of consumption have been reported to increase the risk of strokes caused by bleeding (although it decreases the risk of strokes caused by blocked blood vessels). Impairment of driving skills by alcohol may begin at low blood alcohol concentrations, especially when the drinker is fatigued or taking drugs. Among drugs that enhance alcohol's effects are sleeping pills, tranquillizers, anticonvulsants, antidepressants, and some pain-killers. Small amounts of alcohol may help prevent coronary disease but in severe heart failure, alcohol may only worsen the condition and interfere with the function of medications to treat the disease.

One study indicated that breast cancer was twice more likely to develop in women who drank three to nine drinks per week than in women who drank fewer than three drinks. As noted earlier, the role of alcohol in cancer is controversial, and the breast cancer study was widely challenged.

Finally, adverse affects of alcohol on the fetus have been reported. These effects are usually associated with heavy consumption of alcohol during pregnancy, but not always. One study found that two or three drinks per day during pregnancy was associated with low birth-weight and minor physical anomalies. Another study reported that women drinking two drinks per day during pregnancy had children with lower IQs. Animal research provides additional evidence for adverse fetal effects at low levels of drinking. All of this has resulted in the warning label mentioned earlier that cautions women to avoid *all* alcohol during pregnancy. The Fetal Alcohol Syndrome will be discussed further in Chapter 8.

Finally, moderate drinking does not always remain moderate. Some moderate drinkers progress to heavy drinkers and some become alcoholic, as defined in the next chapter. This book is mainly about alcoholism. However, alcohol-related problems occur frequently in people who are not alcoholic by any definition. For some people there is no such thing as moderate drinking.

SECTION TWO

Alcoholism

In my judgment such of us who have never fallen victims [to alcoholism] have been spared more by the absence of appetite than from any mental or moral superiority over those who have. Indeed, I believe if we take habitual drunkards as a class, their heads and their hearts will bear an advantageous comparison with those of any other class.

—Abraham Lincoln

He drank, not as an epicure, but barbarously, with a speed and dispatch altogether American, as if he were performing a homicidal function, as if he had to kill something inside himself, a worm that would not die.

—Baudelaire, writing about Edgar Allan Poe

5 *What is alcoholism?*

An alcoholic is a person who drinks, has problems from drinking, but goes on drinking anyway:

I am David. I am an alcoholic. I have always been an alcoholic. I will always be an alcoholic. I cannot touch alcohol. It will destroy me. It is like an allergy—not a real allergy—but *like* an allergy.

I had my first drink at sixteen. I got drunk. For several years I drank every week or so with the boys. I didn't always get drunk, but I know now that alcohol affected me differently than other people. I looked forward to the times I knew I could drink. I drank for the glow, the feeling of confidence it gave me. But maybe that's why my friends drank too. They didn't become alcoholics. Alcohol seemed to satisfy some specific need I had, which I can't describe. True, it made me feel good, helped me forget my troubles, but that wasn't it. What was it? I don't know, but I know I liked it, and after a time, I more than liked it, I needed it. Of course, I didn't realize it. It was maybe ten or fifteen years before I realized it, *let* myself realize it.

My need was easy to hide from myself and others (maybe I'm kidding myself about the others). I only associated with people who drank. I married a woman who drank. There were always reasons to drink. I was low, tense, tired, mad, happy. I probably drank as often because I was happy as for any other reason. And occasions for drinking—when drinking was appropriate, expected—were endless. Football games, fishing trips, parties, holidays, birthdays, Christmas, or merely Saturday night. Drinking became interwoven with everything pleasurable—food, sex, social life. When I stopped drinking, these things, for a time, lost all interest for me, they were so tied to drinking. I don't think I will ever enjoy them as much as I did when drinking. But if I had kept drinking, I wouldn't be here to enjoy them. I would be dead.

So, drinking came to dominate my life. By the time I was 25 I was drinking every day, usually before dinner, but sometimes after dinner (if there was a 'reason'), and more on weekends, starting in the afternoon. By 30, I drank all weekend, starting with a beer or Bloody Mary in the morning, and drinking off and on, throughout the day, beer or wine or vodka, indiscriminately. The goal, always, was to maintain a glow, not enough, I hoped, that people would notice, but a glow. When five o'clock came, I thought, well, now it's cocktail hour and I would have my two or three scotches or martinis before dinner as I did on non-weekend nights. After dinner I might nap, but just as often felt a kind of wakeful calm and power and happiness that I've never experienced any other time. These were the dangerous moments. I called friends, boring them with drunken talk; arranged parties; decided impulsively to drive to a bar. In one year, at the age of 33, I had three accidents, all on Saturday night, and was charged with drunken driving once (I kept my licence, but barely). My friends became fewer, reduced to other heavy drinkers and barflies. I fought with my wife, blaming her for *her* drinking, and once or twice hit her (or so she said—like many things I did while drinking, there was no memory afterward).

And by now I was drinking at noontime, with the lunch hour stretching longer and longer. I began taking off whole afternoons, going home potted. I missed mornings at work because of drinking the night before, particularly Monday mornings. And I began drinking weekday mornings to get going. Vodka and orange juice. I thought vodka wouldn't smell (it did). It usually lasted until an early martini luncheon, and I then suffered through until cocktail hour, which came earlier and earlier.

By now I was hooked and knew it, but desperately did not want others to know it. I had been sneaking drinks for years—slipping out to the kitchen during parties and such—but now I began hiding alcohol, in my desk, bedroom, car glove compartment, so it would never be far away, ever. I grew panicky even thinking I might not have alcohol when I needed it, which was just about always.

For years, I drank and had very little hangover, but now the hangovers were gruesome. I felt physically bad—headachy, nauseous, weak—but the mental part was the hardest. I loathed myself. I was waking early and thinking what a mess I was, how I had hurt so many others and myself. The words 'guilty' and 'depression' sound superficial in trying to describe how I felt. The loathing was almost physical—a dead weight that could be lifted in only one way, and that was by having a drink, so I drank, morning after morning. After two or three, my hands were steady, I could hold some breakfast down, and the guilt was gone, or almost.

Despite everything, others knew. There was the odour, the rheumy eyes, and flushed face. There was missing work and not working well when there. Fights with wife, increasingly physical. She kept threatening to leave and finally did. My boss gave me a leave of absence after an embarrassed remark about my 'personal problems'. At some point I was without wife, home, or job. I had nothing to do but drink. The drinking was now steady, days on end. I lost appetite and missed meals (besides, money was short). I awoke at night, sweating and shaking, and had a drink. I awoke in the morning vomiting and had a drink. It couldn't last. My ex-wife found me in my apartment shaking and seeing things, and got me in the hospital. I dried out, left, and went back to drinking. I was hospitalized again, and this time stayed dry for six months. I was nervous and couldn't sleep, but got some of my confidence back and found a part-time job. Then my ex-boss offered my job back and I celebrated by having a drink. The next night I had two drinks. In a month I was drinking as much as ever and again unemployed. That was three years ago. I've had two big drunks since then but don't drink other times. I think about alcohol and miss it. Life is grey and monotonous. The joy and gaiety are gone. But drinking will kill me. I know this and have stopped—for now.

A tree is known by its fruit; alcoholism by its problems. Theoretically, a person can drink a gallon of whisky a day for a lifetime, not have problems, and therefore not be alcoholic. Theoretically. In fact, heavy drinkers almost always have problems. Sometimes they are mild. Alcohol calories may result in overweight—a cosmetic if not a medical problem. Things may be said while drinking that would not or should not be said other times. A minor traffic offence may have major consequences when alcohol is on the breath.

Problems, yes, but alcoholism? The verdict rests with the observer. A fundamentalist teetotaller may view any problem from drinking as alcoholism. Moderate drinkers may be more indulgent, saying in effect, 'These things happen. If they do not happen too often, it probably does not mean much.' But what is too often? Except in extreme cases (the Davids, about whom everyone agrees) there will always be controversy about who is and who is not an alcoholic. This is understandable; doctors disagree about who has heart disease if the case is mild.

'Alcoholism' in this book refers to the David type of alcoholism, granting that patterns of human behaviour are bewilderingly variable, even patterns of illness. Not all Davids, for example, reach bottom (in AA terms). Some stop drinking long before. Others drink, but with enough control to prevent the big

problems from happening. The essence of the David type of
alcoholism is a vulnerability to alcohol that sets him apart from
other drinkers. By taking extreme measures, such as total
abstinence, he may prevent alcohol problems; but if he drinks at
all, the chance of developing problems is high, and this vulnerability appears to be lifelong.

How many people have this condition? It is not known. Population surveys show that about 70 per cent of adults in the USA
drink. About 12 per cent (20 per cent of men, 8 per cent of
women) drink 'heavily', meaning they drink almost daily and
enough to be somewhat intoxicated several times a month.
About 9 per cent have problems from drinking, mostly minor;
another 9 per cent have had problems in the past. (There seems
to be a considerable migration in and out of the 'problem-
drinking' pool.) Among the problem drinkers are a subgroup
called alcoholics. How many alcoholics are there? Nobody
knows, but undoubtedly alcoholics like David exist in large
numbers in all Western countries.

These figures have remained fairly consistent in the USA over
the past 20 years, with, as noted, an overall decline in consumption, at least by non-alcoholic drinkers. It is interesting to compare the situation in the USA with that in the UK. In the late
1980s the World Health Organization sponsored a survey of
nine countries, inquiring about drinking problems. The results
appeared in a report by the Institute of Medicine in the United
States called *Broadening the base of treatment for alcohol problems*
(National Academy Press, Washington, DC, 1990). The report
from the UK came from R. Hodgson of Cardiff. Marcus Grant
and E. B. Ritson, both of the World Health Organization, were
responsible for preparing the summary.

Based on government data, it was estimated that about 10 per
cent of adults in the UK were heavy drinkers. Thirty adults
per 1000 admitted to problems. Five adults per 1000 were problem drinkers. One adult per 2000 was admitted to psychiatric
hospitals with an alcohol-related diagnosis. Most of the problem
drinkers were male. However, medical complications of drinking were increasing faster among women than among men.
Women's deaths from liver disease in the 1980s rose by 9 per

cent compared with a 1 per cent fall for men. There was also a large increase in the number of women seeking alcohol counselling services. The 18–24 age group had the highest proportion of heavy drinkers. There was a marked rise in teenage drunkenness.

Recent data indicate a gradual decline in consumption of alcohol in the UK as in most Western countries. Again, nobody knows how many alcoholics like David exist in the UK. Apparently the number is smaller than in the USA, but, again, this depends on definition.

A disease?

Is alcoholism a disease? The question arises frequently and is the subject of fierce debate among alcoholism experts.

A little historical background may help put the matter in perspective. The 'disease concept' of alcoholism is not new. It originated in the writings of Benjamin Rush and the British physician Thomas Trotter in the early nineteenth century and became increasingly popular with physicians as the century progressed. In the 1830s, Dr Samuel Woodward, the first superintendent of Worcester State Hospital, Massachusetts, and Dr Eli Tood of Hartford, Connecticut, established the first medical institutions for inebriates. *The Journal of Inebriety* was founded in 1876 based on the 'fact that inebriety is a neurosis and psychosis'. In 1904 the Medical Temperance Society changed its name to the American Medical Association for the Study of Inebriety and Narcotics.

The concept of alcoholism as a disease lost favour in the early years of the twentieth century, but came back in vogue in mid-century, in part through pioneering studies at the Yale School of Alcohol Studies and the writings of E. M. Jellinek.

Still, many people resist. Calling alcoholism a disease, they say, simply gives the alcoholic a good alibi for self-indulgence. Maybe a comparison with a 'real' disease would help resolve what is essentially a semantic problem.

Is lead poisoning a disease? Lead poisoning is diagnosed by a specific set of symptoms: abdominal pain, headache, convulsions, coma. Alcoholism also is diagnosed by a specific set of symptoms (reviewed in the next chapter). Both lead poisoning and alcoholism are 'medical' problems, meaning that doctors are supposed to know something about them and possibly be of help.

One reason people, including doctors, have trouble viewing alcoholism as a disease like cancer is that alcoholism is associated with having fun, and fun is not usually associated with disease. (Where does that leave syphilis? Is sex less fun than drinking?)

The point is this: why or how a person 'catches' a disease is not relevant. If some 'self-indulgent' people *enjoyed* lead and ate it like popcorn, this would not change the diagnosis of lead intoxication. Diseases are known by their manifestations as well as their causes, and why alcoholics drink is irrelevant to the diagnosis of alcoholism.

So much for polemics. What has been said above is more or less an American physician's point of view, although widely held in all parts of the world. Still, as mentioned, there is considerable resistance to the disease concept—particularly on the part of psychologists. Opponents believe that 'medicalizing' what is essentially an aberrant form of behaviour—or bad set of habits—is wrong. They point out that millions of people have bad effects from drinking and are not alcoholic by anyone's definition. Nevertheless, they deserve understanding, study, and sometimes professional help. These people, for the purposes of this book, will be called *problem drinkers*, or persons with alcohol-related problems who do not meet criteria for alcoholism. Some of these problems were cited in the previous chapter. The problems may be medical, social, or psychiatric, but they are caused or worsened by alcohol and the person continues to drink *despite* this.

Thus, even with the problem drinker (the person who becomes socially gauche on one drink or suffers excruciating heartburn) the fact that he or she drinks anyway suggests there is an element of *compulsion* in the drinking—just as there is in the most severe form of problem drinking still called, by myself and many others, alcoholism (or sometimes alcohol-dependence).

Alcoholism, in my opinion, is a compulsion to drink that leads to a breakdown in the victim's ability to function. He more than suffers heartburn or social embarrassment. Alcohol, for the alcoholic, is a lethal poison, destroyer of the person's ability to lead anything resembling a normal existence. Alcohol, for the alcoholic, becomes an overpowering obsession that leads to a compulsion to drink that is so strong it dominates a person's life. Obtaining the next drink becomes a need that obliterates every other aspect of a person's psychological being and the alcoholic's body comes to depend on alcohol almost as much as it depends on oxygen and food.

This is my working definition of alcoholism. When the term is used in this book, that is what I am talking about. David, in my opinion, was an alcoholic.

The critics of the disease concept are right in one regard. People have problems in drinking who are not alcoholic by the above definition and never will be. The non-alcoholic problem drinker is also the subject of this book.

6 *The symptoms*

This chapter is mainly about alcoholics, heavy drinkers for whom drinking has become a central activity in their way of life. But many non-alcoholics have experienced some of the problems, addressed below, and these will be discussed at the end of the chapter.

Psychological problems

The symptoms of alcoholism fall into three groups: psychological, medical, and social. Let me start with those that are psychological.

1. *Preoccupation with alcohol*

The alcoholic thinks about alcohol from morning till night, and at night, if not too drunk to dream, dreams about alcohol. When to have the first drink? When the next? Remember the times that bars will be closed. Remember liquor stores are closed on Sundays. Prepare, prepare. Will they sell more than two drinks on the aeroplane? Take a flask. Do the Smiths drink? Find out before accepting their dinner invitation. This goes on and on, blotting out other thoughts, other plans.

It is obsessional in precisely the way psychiatrists use the word. Obsessions breed compulsions, and when an alcoholic drops in on a bar or liquor store, ever so casually, it is as compulsive as the neurotic washing his hands for the twentieth time that day.

2. *Self-deception*

But he must not admit it to himself. 'We are all victims of systematic self-deception,' Santayana said, and the alcoholic is a victim *par excellence*. People are victims of many things—cancer, lust, society—and can accept it. But, deep down, the alcoholic believes he is doing it to himself; he is the perpetrator, not the victim. And this he cannot accept, so he lies to himself.

'I can stop drinking anytime. Important people drink. Churchill drank. Today is special—a friend is in town. Nothing is going on—why not? Life is tragic—why not? Tomorrow we die—why not?'

As he lies to himself, he lies to others, and concealment becomes a game like the one children play when they raid the cookie jar and hope their mother won't notice.

3. Guilt

But he does know and can't help knowing. There are too many reminders. The wife's pleas and tantrums. The boss's 'friendly' advice. The dented bumper. The night terrors and night sweats. The trembling hands. The puffy eyes and blotchy complexion. The terrifying memory gaps. All spell self-destruction, and even the cleverest self-deceiver knows it.

4. Amnesia

Alcoholics have memory lapses when they drink and this is often attributed to guilt. It is said the forgetter does not want to remember. Non-alcoholics also have memory lapses when they drink, not so often or so severely, but non-alcoholics by definition drink less. Memory lapses (or blackouts, as they are called when alcohol is involved) are probably not due to guilt. More likely alcohol, in some people on some occasions, interferes with chemical processes that make memory (perhaps the most mysterious of biological phenomena) possible.

Precisely how it occurs is unknown, but the memory lapses are genuine. The drinker does things when he is drinking that ordinarily he would remember perfectly, but when he sobers up, usually the next day, he has no recollection of what he has done. Sometimes he realizes that he had a memory lapse. He is apprehensive. He checks to see if his car is in the garage. He

looks for dents that weren't there before. His overriding fear is
that he did something—broke a law, harmed someone—and
punishment is at hand. He retraces his movements of the night
before. 'Was I here, Joe?' he asks. Told that he was: 'What did
I do? Was I drunk?' Reassured that he did nothing wrong and
was no more drunk than usual, he goes to the next place where
he might have been. Alternately, he may avoid all places and all
companions he might have visited or been with during the
forgotten interval, preferring not to know.

In truth, people rarely do things during blackouts that they
don't also do when they are drunk and suffer no memory loss.

During blackouts, the person is conscious and alert. He may
appear normal. He may do complicated things—converse intel-
ligently, seduce strangers, travel. A true story:

A 39-year-old salesman awoke in a strange hotel room. He had a mild
hangover but otherwise felt normal. His clothes were hanging in the
closet; he was clean-shaven. He dressed and went down to the lobby.
He learned from the clerk that he was in Las Vegas and that he had
checked in two days previously. It had been obvious that he had been
drinking, the clerk said, but he hadn't seemed very drunk. The date
was Saturday the 14th. His last recollection was of sitting in a St. Louis
bar on Monday the 9th. He had been drinking all day and was drunk,
but could remember everything perfectly until about 3 p.m., when 'like
a curtain dropping', his memory went blank. It remained blank for
approximately five days. Three years later, it was still blank. He was so
frightened by the experience that he abstained from alcohol for two
years.

Some people forget and do not realize when sober that they
have forgotten anything. Someone tells them and then they
remember a little.

A 53-year-old member of Alcoholics Anonymous said that he had
experienced many blackouts during his 25 years of heavy drinking. He
could not remember his first blackout, but guessed it had happened
about 15 years before. The memory loss had not bothered him, he said;
he assumed everyone who drank had trouble with memory. Sometimes, however, it was embarrassing to be told he had said something
or gone somewhere and not recalled it. Upon being told, he would
sometimes remember the event and sometimes not. Occasionally,

months later, something would remind him of the event and his memory would 'snap back'. Typically, he could remember some parts of a drinking episode and not others; a half hour might be blanked out and the next hour remembered. The forgotten parts appeared to have no more emotional significance than the remembered ones. 'It's like turning a switch on and off.'

Sometimes a curious thing happens when a person is drinking: the drinker recalls things that happened during a previous drinking period which, when sober, he had forgotten. For example, alcoholics often report hiding money or alcohol when drinking, forgetting it when sober, and having their memory return when drinking again. This is reminiscent of the 'state-dependent learning' described in the first chapter. Whatever the explanation, the mind does play odd tricks on drinkers:

A 47-year-old housewife often wrote letters when she was drinking. Sometimes she would jot down notes for a letter and start writing it but not finish it. The next day, sober, she would be unable to decipher the notes. Then she would start drinking again, and after a few drinks the meaning of the notes would become clear and she would resume writing the letter. 'It was like picking up the pencil where I had left off.'

5. *Anxiety and depression*

What goes up comes down, and alcoholic euphoria is followed by alcoholic depression with a kind of Newtonian inevitability. Anxiety and depression occur not only with hangovers but intermittently during the drinking period itself, if the drinking is heavy and continuous. This sequence is common: a man feels bad for any reason (it's a grey day); he drinks, feels better; then he feels bad again, this time because the alcohol effect is wearing off; he drinks again, feels better again. And a vicious circle is under way, based on alcohol's ability to alternately raise and lower spirits. This rollercoaster effect is probably chemical in nature, but the drinker only knows that alcohol, having raised his spirits, now lowers them, and that the best way to raise them again is to have another drink.

Medical problems

If this process goes on long enough, there are usually medical problems. They may take years to develop, and some lucky drinkers never are affected. Incredibly, a person may consume a ferocious quantity of alcohol, maybe a fifth or a quart of whisky a day for twenty years or longer, and when he dies a 'natural' death, his brain, liver, pancreas, and coronary arteries appear normal. But the odds are strong that something will give. Here are some favourite targets.

1. *The stomach*

Gastritis, inflammation of the stomach's lining, is common. The symptoms are gas, bloating, heartburn, nausea. The cure is alcohol. Before-and-after pictures of the stomach prove it. Raw and inflamed from a night, or week, of heavy drinking, the stomach is miraculously restored to a normal appearance after a shot or two of alcohol.

Alcohol's role in producing stomach or duodenal ulcers is debatable. Ulcers are caused by hydrochloric acid and digestive enzymes. These are powerful enough to digest fish bones and the toughest beefsteak, but inexplicably do not digest the stomach itself. The mucous lining of the stomach somehow prevents it. When the protection is lost, ulcers develop. If alcohol, hot peppers, or pent-up anger produce ulcers at all, they must do so either by increasing acid production or by breaching the protective barrier. But alcohol, in strong doses, decreases rather than increases acid production. Its effect on the protective lining is not clear.

2. *The liver*

The word 'cirrhosis' comes from the Greek word for yellow-orange, probably because people with cirrhosis become jaundiced. Alcoholics are disposed to a type of cirrhosis called portal cirrhosis, or Laennec's cirrhosis.

In the first stages of cirrhosis, liver cells become inflamed and gradually die out. The liver swells up and can be felt through the belly wall, whereas usually it is hidden behind the ribs. New cells appear (the body incorrigibly bent on keeping things going) but with a difference. Previously the cells lined up in columns, forming banks for concentric canals through which blood coursed. The new cells form higgledy-piggledy, and the blood flow comes nearly to a halt.

The results are predictable. Blood backs up and seeps into the abdomen, which swells like a balloon, or it detours around the liver, engorging the paper-thin veins of the oesophagus. If the veins burst, fatal haemorrhage may result. The liver cells, formerly little factories with many functions, go on strike and their production of proteins, blood-clotting factors, and other vital constituents falls off. In men, the cells no longer suppress female sex hormones (the manliest man has *some* female sex hormones), so men's breasts grow, their testicles shrink, and they lose their baritone voices, beards, and hairy chests.

As the process continues, scar tissue forms and eventually the liver is like a small lumpy rock, incapable of sustaining life.

Cirrhosis is a leading cause of death in Western countries, and most people with Laennec's cirrhosis are alcoholics. Most alcoholics, however, do not develop cirrhosis, and the connection between drinking and cirrhosis is still not understood.

The risk of liver disease increases when certain drugs are taken together. Carbon tetrachloride and other halogenated hydrocarbons can produce liver damage alone, but when they are combined with alcohol the risk of liver damage is much greater. Acute and sometimes catastrophic liver disease has occurred in individuals who devote a Saturday afternoon to scrubbing their wall-to-wall carpets with a cleaning fluid while drinking half a case of beer. Acetaminophen, a widely used aspirin substitute, may produce serious liver damage when combined with alcohol over a period of time.

3. *Nerve fibres*

The long nerve fibres extending from the spinal cord to muscles often suffer degenerative changes in alcoholics. The fibres make muscles contract and maintain muscular tone; they also transmit back to the spinal cord, and thence the brain, messages from sensory receptors in muscle and skin. The degeneration of nerves results in muscular weakness and eventual wasting and paralysis. Pain and tingling are experienced, and there may be eventual loss of sensation. The cause of nerve-fibre degeneration in the alcoholic is not alcohol; it is a vitamin deficiency. High doses of B vitamins almost always restore the fibres to their normal state, if not given too late.

4. Brain damage

After many years of heavy drinking, most alcoholics, when recovered from their latest drinking bout, show little or no sign of intellectual impairment. Their IQs are normal, their thoughts logical, and their minds clear. If there is impairment, it is usually subtle, rarely persists, and can be attributed to factors other than loss of brain cells—poor motivation, for example, in taking the tests psychologists are forever giving alcoholics.

Imaging techniques, including computerized X-rays of the brain (CAT scans), have been applied to alcoholics in several studies. The results have been mixed. Some studies show a loss of brain tissue in alcoholics, others show no loss, and still others show loss followed by a return to normal! Rats, in one recent study, suffered loss of brain cells after prolonged alcohol intoxication. Intriguingly, the loss occurred mainly in regions of the brain associated with memory processes. In view of the frequent blackouts experienced by alcoholics, it is tempting to speculate that alcohol has a special affinity for these areas of the brain, but more study is needed to prove this. Magnetic nuclear imaging (MRI) studies also show loss of brain tissue in alcoholics, but, again, it is not clear whether the loss is genuine and permanent or possibly artifacts caused by fluid shifts in the brain.

A small minority of alcoholics definitely suffer brain damage due to deficiency of thiamine, a B vitamin. The malnourished alcoholic gets too little thiamine, and if the deprivation persists and is severe, certain well-demarcated areas of the brain are destroyed. These areas are definitely involved in memory storage. Their destruction results in severe memory impairment. A German named Wernicke and a Russian named Korsakoff first described the disease. The patient with Wernicke–Korsakoff disease can remember the distant past fairly well, has a normal IQ, and seems reasonably bright; however he is unable to remember anything that happened to him a few minutes after it happens. The condition is devastating, and the chronic Wernicke–Korsakoff patient needs custodial care for the rest of his life. Thiamine, given early, may prevent a permanent defect. Fortunately, the condition is rare.

Alcoholics also are inclined to suffer degenerative changes in the cerebellum, the half-melon bulge at the base of the brain that regulates co-ordination. An unsteady gait results. Vitamin deficiency is believed to be the cause.

5. *Impotency*

MACDUFF: What three things does drink especially provoke?
PORTER: Marry, sir, nose-painting, sleep and urine. Lechery, sir, it
          provokes, and unprovokes; it provokes the desire, but it
          takes away the performance. . .

Macbeth (Act II, scene iii)

Shakespeare names not three but four of alcohol's well-known actions, lechery being the most famous. By dilating blood vessels, alcohol 'paints' the nose; it makes people sleepy; and one of the things a novice drinker first notices about drink, it increases the need to urinate.

It also increases sexual desire. More accurately, perhaps, it 'releases' sexual desire—the well-known disinhibiting effect of alcohol. But performance may be impaired. Drunken men have trouble achieving an erection or ejaculating. Whether sexual performance in drunken women also is impaired is hard to determine. Women have less to erect, and the female orgasm remains poorly understood.

Some alcoholics not only have trouble with sexual perform-ance when drinking; the problem persists long into sobriety.

Whether the cause is psychological or physical is not known, but it may contribute to a well-described syndrome: alcoholic conjugal paranoia. Husbands, without evidence, become convinced that their wives are unfaithful. They hound their wives, accuse them, search for anything to support their delusion: inspect underwear for semen spots, hire detectives, sniff blouses for aftershave lotion. The delusion is precisely that: a fixed false idea. It may be related to the husband's feelings of inadequacy about his own sexual ability and perhaps to feelings of inadequacy about life in general, in ruins from years of heavy drinking. Whether women alcoholics also develop alcoholic paranoia is not clear.

6. *Other medical problems*

'To know syphilis is to know medicine', Wiliam Osler said at a time when syphilis was untreatable and affected, or could affect, nearly every organ in the body. The same can be said of alcoholism. It can affect, or is alleged to affect, nearly every organ. Rare is the medical journal that does not occasionally publish new evidence of alcohol's dangers: heart disease, muscle disease, pancreatitis, anaemia, cancer, not to mention the conditions described above. The list is long and growing. But are the reports reliable and is the culprit alcohol?

The problem in ascribing an illness to heavy drinking is that heavy drinkers differ from non-heavy drinkers in other ways. They smoke more. They often eat less. They often lead irregular lives—staying up all hours, never exercising, sleeping it off on park benches. How can these potentially harmful influences be separated from the effects of alcohol? It is difficult.

Moreover, the reports associating alcohol with a particular illness often are contradictory. At least one study reports that alcoholics have less heart disease, not more (see Chapter 4). Another claims that drinkers live longer than teetotallers, but the 'drinkers' studied were probably not alcoholics.

In conclusion, it is a fact that serious medical problems are associated with alcoholism, although why, how, and how many, remains unknown.

The alcohol-withdrawal syndrome

Alcoholics also experience a medical problem that, strictly
speaking, does not come from drinking alcohol but from *not*
drinking alcohol. This is the alcohol-withdrawal syndrome. It is
commonly, but mistakenly, called the DTs, or delirium tremens.
In medical usage 'delirium' means gross memory disturbance,
usually combined with insomnia, agitation, hallucinations, and
illusions. Most alcoholics do not experience delirium.

As a rule, alcohol withdrawal is a distressing but temporary
condition lasting from two days to a week. The mildest symptom
is shakiness, which begins a few hours after the patient stops
drinking, sometimes awakening him during sleep. Morning
shakes are inevitable if the drinker has been drinking enough.
His eyelids flutter, his tongue quivers, but, most conspicuously,
his hands shake, so that transporting a cup of coffee from saucer
to mouth is a major undertaking.

The cure for the shakes, as for all alcohol-withdrawal symptoms, is a drink or two.

After a day or two without drinking, the alcoholic coming off a
bender may start hallucinating—seeing and hearing things that
others do not see or hear. He often realizes he is hallucinating
and blames alcohol. Not always, however. Sometimes the hallucinations are vivid, frightening, and as real as life.

Occasionally alcoholics have convulsions that resemble the
grand mal seizures of the epileptic. Most alcoholics are not epileptic and have seizures only when withdrawing from alcohol.
The seizures usually occur one to three days after the person
stops drinking.

The most severe form of withdrawal involves delirium and
justifies using the term delirium tremens, the 'tremens' referring to the shakiness. Delirium is ominous. It often means the
person has not only withdrawal symptoms but also a serious
medical illness, often of the type to which alcoholics, because
of their way of living, are vulnerable: pneumonia, fractures,
blood clots in the brain, liver failure. People occasionally die in
delirium tremens, whereas death from milder forms of withdrawal is rare.

Many alcoholics are capable of withdrawing from alcohol on their own. They often do this by tapering off—gradually decreasing the amount they drink. Serious withdrawal symptoms, however, justify hospitalization so that tranquillizers can be given to make the alcoholic feel better, vitamins to prevent brain damage, and frequent medical examinations to exclude medical illness.

The DTs have been described brilliantly in fiction by among others, Malcolm Lowry and Mark Twain. Lowry, in his novella 'Lunar caustic', wrote from personal experience how it felt to wake up in an alcoholic ward:*

The man awoke certain that he was on a ship. If not, where did those isolated clangings come from, those sounds of iron on iron? He recognized the crunch of water pouring over the scuttle, the heavy tramp of feet on the deck above, the steady Frère *Jac*ques: Frère *Jac*ques of the engines. He was on a ship, taking him back to England, which he never should have left in the first place. Now he was conscious of his racked, trembling, malodorous body. Daylight sent probes of agony against his eyelids. Opening them, he saw three negro sailors vigorously washing down the deck. He shut his eyes again. Impossible, he thought . . .

As day grew, the noise became more ghastly: what sounded like a railway seemed to be running just over the ceiling. Another night came. The noise grew worse and, stranger yet, the crew kept multiplying. More and more men, bruised, wounded, and always drunk, were hurled down the alley by petty officers to lie face downward, screaming or suddenly asleep on their hard bunks.

He was awake. What had he done last night? Nothing at all, perhaps, yet remorse tore at his vitals. He needed a drink desperately. He did not know whether his eyes were closed or open. Horrid shapes plunged out of the blankness, gibbering, rubbing their bristles against his face, but he couldn't move. Something had got under his bed too, a bear that kept trying to get up. Voices, a prosopopoeia of voices, murmured in his ears, ebbed away, murmured again, cackled, shrieked, cajoled; voices pleading with him to stop drinking, to die and be damned. Thronged, dreadful shadows came close, were snatched away. A cataract of water was pouring through the wall, filling the room. A red hand gesticulated, prodded him: over a ravaged mountain side a swift stream

* 'Lunar caustic' appeared in the *Paris Review* in the Winter–Spring issue of 1963.

was carrying with it legless bodies yelling out of great eye-sockets, in which were broken teeth. Music mounted to a screech, subsided. On a tumbled bloodstained bed in a house whose face was blasted away a large scorpion was gravely raping a one-armed negress. His wife appeared, tears streaming down her face, pitying, only to be instantly transformed into Richard III, who sprang forward to smother him.

After a few days, the DTs go away. Lowry's patient 'now knew himself to be in a kind of hospital, and with this realization everything became coherent and fell into place. The sound of water pouring over the scuttle was the terrific shock of the flushing toilets; the banging of iron and the dispersed noises, the rattling of keys, explained themselves; the frantic ringing of bells was for doctors or nurses; and all the shouting, shuffling, creaking and ordering was no more than the complex routine of the institution.'

Psychiatric patients are rarely dangerous, but delirious patients are an exception. They may be dangerous indeed, as was the case of Huckleberry Finn's alcoholic father, whose DTs were described by Mark Twain as follows:

I don't know how long I was asleep, but all of a sudden there was an awful scream and I was up. There was Pap looking wild, and skipping around and yelling about snakes. I couldn't see no snakes, but he said they was crawling up his legs; and then he would give a jump and scream, and say one had bit him on the cheek. I never see a man look so wild. Pretty soon he was all fagged out, and fell down panting; then he rolled over and over, screaming and saying there was devils a-hold of him. He wore out by and by, and laid still awhile, moaning. Then he laid stiller, and didn't make a sound. I could hear the owls and the wolves away off in the woods, and it seemed terrible still. He was laying over by the corner. By and by he raised up partway and listened, with his head to one side. He wails, very low:

'Tramp-tramp-tramp; that's the dead; tramp-tramp-tramp; they're coming after me; but I won't go. Oh, they're here! Don't touch me— don't. Hands off—they're cold; let go. Oh, let a poor devil alone!'

He rolled himself up in his blanket and went to crying. But by and by he rolled out and jumped up to his feet looking wild, and he see me and went for me. He chased me round and round the place with a clasp knife, calling me the Angel of Death, and saying he would kill me . . . I begged, and told him I was only Huck; but he laughed such a screech laugh, and roared and cussed, and kept on chasing me. Once when I turned short and dodged under his arm he got me by the jacket between my shoulders, and I thought I was gone; but I slid out of the jacket and saved myself. Pretty soon he was tired out, and dropped down with his back against the door, and said he would rest a minute and then kill me. He put his knife under him, and pretty soon he dozed off.

Social problems

In simpler times, it was said that marijuana smoking was a 'crime without a victim', but even then no one would have called alcoholism a victimless 'crime'. The victims of alcoholism are legion: spouses, children, other relatives, bosses, fellow workers, pedestrians, drivers, police, judges, physicians who get called late at night, taxpayers who often pick up the bill for treatment, and other innocent and not so innocent people who cross the alcoholic's path. Here are some telling statistics.

1. Before drugs became a major problem in the USA and other countries, the average city policeman spent one-half of his time dealing with alcohol-related offences. Nearly half of the men and women in prison were alcoholic or, at any rate, heavy drinkers. Most murderers were drinking at the time they committed a murder, and so were most of the victims, although how many would be considered alcoholic is uncertain. Now US jails and prisons are full of people arrested for crimes involving illicit drugs. Many of them are also heavy drinkers. The relative contribution of drugs and alcohol to their criminal activities becomes difficult to parcel out. This also applies to the problems mentioned below.

2. Between 20 and 30 per cent of male psychiatric admissions are alcoholic or have alcohol-related problems. About one-quarter of the men admitted to general hospital wards for medical treatment have alcohol-related problems.

3. Industry loses huge amounts of money every year because of absenteeism and work inefficiency related to alcoholism. Monday morning and Friday afternoon absenteeism, at least

partly attributable to alcoholism, is so common that both industry and unions are considering a four-day working week (whereupon Tuesday morning or Thursday afternoon absenteeism will probably become common).

4. Alcoholics are about two times more likely to be divorced than non-alcoholics.

5. Alcoholics have a death rate at least two times higher than non-alcoholics. The most common causes, aside from medical diseases, are accidents and suicides. There are an estimated 20 000 deaths a year in the USA from alcohol-related automobile accidents. Studies indicate that most of the drinking drivers are not just social drinkers coming home from a Christmas party but serious problem drinkers, alcoholic by most definitions. About one out of four suicides in the USA is an alcoholic, usually a man over 35.

Comparable figures are not available for the UK. However, a WHO report cited the social costs of alcohol use (using 1983 prices) as being 1600 million pounds per year, and this was considered a low estimate. The figure included cost to industry, cost to the National Health Service, and the social cost of drink-related offences. (One reason why it is hard to compare US and UK statistics is that the USA still uses the term alcoholism freely, while the UK speaks about 'drink-related offences'.) It includes the cost to society of family stress and the burden placed on probation and social services. Days lost from work because of alcohol-induced sickness were estimated at 8 million to 15 million days per year in the UK. Of male admissions to a general medical ward, 20 per cent were estimated to be related to alcohol use.

The UK report said that alcohol intoxication was involved in 60 per cent of suicide attempts, 54 per cent of fire fatalities, 50 per cent of homicides, 42 per cent of patients with serious head injuries, 30 per cent of deaths through drowning, and 30 per cent of domestic accidents. Alcohol intoxication was 'implicated' in the deaths of more than 500 young people each year, about 10 per cent of deaths in persons under 15. Alcohol usage was associated with assaults in 78 per cent of cases and breaches of the peace in 80 per cent of cases. Furthermore,

93 per cent of people arrested between the hours of 10 p.m. and 2 a.m. were intoxicated. About 1 in 3 drivers killed in road accidents had blood alcohol levels about the statutory limit. On Saturday night this figure rose to 71 per cent.

These figures were based on government data from the 1970s through the mid-1980s. They may have changed somewhat since then, and may indeed be somewhat lower, but they are still in the 'same ball park' (to use an American expression).

Alcohol-related problems

In one survey of American drinkers, 18 per cent had experienced at least one problem from drinking in their lifetime. They also, at some point in their life, were heavy drinkers, meaning they drank almost every day and six or more drinks on at least one occasion per week. Half of the heavy drinkers (9 per cent) stopped being heavy drinkers for several years and had no further difficulty with alcohol.

Should these former heavy drinkers who later drank normally be called alcoholics? Whether to do so is partly a matter of definition but also reflects often intense, almost religious views of what alcoholism is and is not. There is no question that many people who have *minor* and *infrequent* problems from drinking can control their drinking. An alcoholic, by definition, is a person with *severe* alcohol problems who cannot control his drinking. More about this later.

Following the order in which problems for the alcoholic were described, here are some comments about these problems as experienced by non-alcoholics.

It is not true to say that non-alcoholics do not think much about alcohol. Social drinkers (as they are sometimes called) may have excellent control over the amount they drink and still look forward to a preprandial martini (and miss it when they do not get one). They look forward to what used to be called 'getting drunk and having fun' on New Year's Eve (this is now considered a tactless remark in many segments of US society). Many college students look forward to the beers they plan to consume after taking a tough examination. It is still socially acceptable, in most parts of society, to celebrate weddings, birthdays, anniversaries, and reunions with an old friend with a drink (often not limited to one drink). George Simenon had this to say about drinking in America (and he was not talking solely about alcoholics): '. . . [Americans experience] an almost permanent state in which one is dominated by alcohol, whether during the hours one is drinking or during the hours when one is impatiently waiting to drink, almost as painfully as a drug addict waits for his injection . . . If one has never known this experience, it is difficult to understand American life. Not that everyone drinks, in the sense in which my mother used the word, but because it is part of private and public life, of folklore, you might say, as is proved by the large, more or less untranslatable vocabulary, most often in slang, that relates to drink . . .'

'All of life is colored by it. New York, for example, seems made to be seen in this state, and then it is an extraordinary New York and, strange as it may seem, comradely.'

'The crowds cease to be anonymous, the bars cease to be ordinary ill-lit places, the taxi-drivers complaining or menacing people. It is the same for all the big American cities. Los Angeles, San Francisco, Boston . . . From one end of the country to the other there exists a freemasonry of alcoholics . . .' Simenon's description may have fitted America better in the 1920s and 1950s (when he wrote it) than it does in the 1990s. Attitudes toward drinking have changed in the past thirty years, with drunkenness much less tolerated. Still, bars continue to outnumber churches (even grocery stores) in American cities, despite the legal perils of embarking on the highway after drinking as little as two drinks (the legal limit for intoxication has been dropped to 0.08 per cent in some American states, which two drinks will produce in some people). Beer commercials continue to dominate sports commercials, even if the beer today is sometimes non-alcoholic beer. Many Americans are at least somewhat preoccupied with alcohol, even though they would not be described as alcoholic.

The non-alcoholic heavy drinker also often lies to himself—or at least to his doctor. Physicians are convinced that most drinkers minimize the amount of alcohol they drink when asked, often with regard to controlling their weight or rule out a possible cause for a medical condition. People who drink even moderately may have concern that their drinking may not be so moderate after all—that it may indeed lead to serious drinking—but they do not want to give up all alcohol if they can help it, so they minimize the amount they drink. (In alcoholics, this would be called 'denial'.)

Does the moderate drinker with an occasional problem from drinking experience guilt? Sometimes. Simenon said, 'In the United States I learned shame. For they are ashamed. Everyone is ashamed. I was ashamed like the rest.' Any drinker who has been charged with drunken driving because he had three or four drinks on one occasion (perhaps more than he had ever consumed previously) knows the feeling. Anyone who has had a few drinks at an office party and sexually harassed his secretary knows the feeling. Anyone who has had a hangover probably knows the feeling. Kingsley Amis, the British novelist, said hangovers take two forms: physical and metaphysical. He considered the metaphysical part (the guilt) far more distressing than nausea and headache.

More than a third of middle-class American men have experienced at least one alcoholic blackout (as defined above) by the age of twenty. It has been said that having a memory lapse (blackout) while drinking as a young person means the person is doomed to become alcoholic later in life. This has been shown to be untrue. Most people who have experienced one or even several blackouts do *not* become alcoholics. On the other hand, alcoholics have *many* blackouts, particularly as they get older.

For some people, even a single drink will make them feel depressed. After several drinks they may feel profoundly depressed, especially the next morning. This has been called *alcohol-induced-dysphoria*. It only happens to a minority of people, but it is one reason people who are not destined to become alcoholic give up alcohol entirely. It just isn't worth it.

Incidentally, alcohol-induced dysphoria appears to occur most often in people who have a susceptibility to depression when not drinking. Manic-depressives, for example, tend to drink *less* when depressed than when not depressed, probably because alcohol loses its interest for people who are depressed or they have found that even a little alcohol makes them feel more depressed, particularly the next morning.

As for medical problems from drinking, even small amounts of alcohol will cause heartburn or diarrhoea in some people, and may worsen a stomach or duodenal ulcer if one is already present. Drinkers who develop a particular type of cirrhosis are usually described as alcoholic but there is some disagreement about this. Some studies have reported cirrhosis occurring in people who staunchly defend themselves as moderate drinkers. Doctors' scepticism about drinking histories, born of hard experience, may result in an over-diagnosis of alcoholism in moderate drinkers.

Pancreatitis can certainly occur in drinkers who consume far less than does the typical alcoholic.

Some experts believe even modest amounts of alcohol consumed on a regular basis may cause subtle impairment of brain functioning that can become permanent. Most experts believe that drinkers who limit their alcohol to a drink or two per day are probably in no danger of brain damage.

Moderate drinkers who drink immoderately on some occasion are as susceptible to sexual dysfunction as the hard-drinking alcoholic.

There is some evidence that moderate drinking may promote cancer of the breast or cancer of the rectum and colon. More evidence is needed before this can be viewed as conclusive.

Even light drinkers often believe that a drink or two at night takes the edge off their effectiveness the next day or may contribute to migraine headache (which it probably does) or a downright unwillingness to show up for work, at least until noon. This may be all in the imagination, as the saying goes, but many people believe it.

Finally, moderate use of alcohol in some drinkers may produce social problems. Some men may physically abuse their wives or children who would never think of doing so if they had not had two or three drinks (usually it takes more). Two or three drinks may produce just the amount of courage needed to rob a convenience store. Two or three drinks may not mix well with prescription tranquillizers and sleeping pills or non-prescription cold medication. These combinations may produce driving errors and social bad judgments that would not have occurred from one ingredient alone.

7 The course

Disease, again

Diseases tend to have a predicable course or 'natural history'—
to progress or terminate in more or less predictable ways if not
modified by treatment. In the USA alcoholism is widely considered a disease; the American Medical Association has taken
this position. Others disagree—especially in the UK. Many opponents of the disease concept believe 'problem drinking' or
'heavy drinking' (they avoid the medically and AA-tainted
word alcoholism) is a form of learned behaviour and believe it
should be treated, not as a disease with drugs or even psychotherapy, but as a learned behaviour that can be *unlearned* by
appropriate techniques, usually applied by psychologists. They
accept the idea that dependence (physical and psychological)
occurs, but believe there are *degrees of dependence* rather than an
all-or-nothing phenomenon called alcoholism. When talking
about *extreme* dependence, they describe something close to
what is called alcoholism, but avoid the word. They believe even
extremely dependent drinkers can be treated by modifying what
is basically learned behaviour. They are distressed when heavy
drinkers are viewed as 'helpless victims of a disease', which
may only serve as an excuse for continued drinking. This point
of view is cogently presented in a book by Hubert Fingarette
called *Heavy drinking—the myth of alcoholism as a disease* (University of California Press, 1988).

It is with regard to the *course* of the illness (which opponents
do not consider an illness) or the *natural history* that one finds
the strongest disagreement. Advocates of the learning theory
approach believe even severely dependent drinkers can learn
to drink normally, at least in many cases. Here the disease–
non-disease conflict clashes head on. Extremely dependent
drinkers, called alcoholics by most American therapists, are
admonished that they can *never* drink normally, and to try ends

ultimately in disaster. Their belief that total abstinence from alcohol is the only practical goal of treatment is based in part on the philosophy of Alcoholics Anonymous and is also held by most American physicians and physicians in other countries.

To a large extent this is a straw-man argument. There is no controversy about whether most non-dependent 'heavy drinkers' could drink in moderation if they decided to do so, and surely no one would call heavy drinking in itself a disease. What is controversial is whether severely dependent heavy drinkers—alcoholics—can often return to normal drinking. For non-dependent drinkers, it is true: there is no natural history or predictable course; many go on to become controlled problem-free drinkers with or without professional help. Those who do not succeed in this, despite professional help, are usually the most dependent drinkers.

Concerning the word 'dependence': it refers to physical dependence—where a person must use alcohol to avoid withdrawal symptoms—or psychological dependence, where the person's life revolves around alcohol. Whether there is a continuum between the non-dependent heavy drinker and the severely dependent drinker is not known. There is evidence, in fact, that 'true' alcoholism is *not* on a continuum with heavy drinking but a separate entity, but this will be discussed later.

Finally, the disease debate is basically a semantic argument revolving around the *definition* of disease. Until debaters agree on the definition, they cannot possibly hold a useful debate. ('Words', as Humpty Dumpty said, 'mean just what I chose them to mean—neither more nor less.')

For definitions we often turn to dictionaries. This is not very helpful in regard to disease. In many dictionaries disease is defined as 'absence of health', just as health is defined as 'absence of disease'.

In truth, there are three useful definitions of disease: narrow, broad, and clinical.

The broad definition is this: diseases are what people go to doctors for. They presumably do so because they believe doctors can help them (whether this is true or not). What people go to doctors for changes over time. A hundred years ago, bank robbers and child molesters saw judges—today they may see a psychiatrist as well. Alcoholics see doctors, sometimes for alcoholism.

The term 'disease' also suggests suffering and sometimes death, and surely no one would deny that these are attributes of alcoholism.

The narrow definition of disease holds that it involves a biochemical, physiological, or anatomical abnormality, the nature of which may not be known.

Alcoholism, in fact, may even meet the narrow definition. Alcoholism runs in families, even when the children are separated from their alcoholic parents and raised by non-alcoholic adoptive parents. This suggests a biological susceptibility (or abnormality if you prefer). Twin studies also indicate a genetic factor in alcoholism.

The clinical definition of disease goes as follows: a disease is a cluster of symptoms and/or signs with a more or less predictable course. Symptoms are what patients tell you; signs are what you see. A cluster may be associated with physical abnormality or may not.

In fact, a disease is a category used by physicians, as 'apples' is a category used by grocers. It is a useful category if precise and if the encompassed phenomena are stable over time. Diseases are conventions and may not 'fit' anything in nature at all. Through the centuries diseases have come and gone, some more useful than others, and there is no guarantee that our present 'diseases' (medical or psychiatric) will represent the same clusters of symptoms and signs a hundred years from now that they do today. On the contrary, as more is learned, more useful clusters surely will emerge.

Disease, as defined, is synonymous with the so-called 'medical model' and it is true that medical doctors tend to view disease in this manner. However, this does not limit the recognition and treatment of alcohol dependence to physicians. At present in the USA, non-physicians (many called alcoholism counsellors) are the primary treaters of alcohol problems. They usually accept the 'medical model' and disease concept. (The UK approach is discussed in Chapter 17.)

The clinical disease concept is the basis for this chapter because of its emphasis on natural history. Alcohol-dependence, in its severest form, does have a natural history. What follows is about severely dependent drinkers and not about people with alcohol-related problems who do not fit the definition of alcoholism.

Natural history

When does alcoholism begin? Fixing the onset of a chronic condition is difficult. In cancer, when does the first cell become malignant? When does the first coronary artery narrow in heart disease? Cancer and heart diseases usually can be diagnosed only after they are far advanced, and the same is true of alcoholism. With some alcoholics, the alcoholism seems to start with the first drink, but if this happens, often it is not apparent except in retrospect.

Nevertheless, the condition tends to develop at certain ages, progresses in a more or less predictable manner, and terminates in more or less predictable ways. 'More or less' is an important qualifier, as there is much variation. Men and women vary; whites and blacks; Americans and French.

The 'typical' white male alcoholic begins drinking heavily in his late teens or early twenties, drinks more and more throughout his twenties, starts having serious problems in his thirties, is hospitalized for drinking (if ever) in his mid- or late-thirties, is clearly identified by himself and others as alcoholic—a man who cannot drink without trouble—between the ages of 40 and 50.

Men, with rare exceptions, do not become alcoholic after 45. There is an 'age of risk' for alcoholism, as for most illnesses, and if a man has no symptoms of alcoholism by his late-forties, he probably will develop none.

The illness ends by death from suicide, accident, or medical illness—or by cessation of drinking. Few alcoholics, in the view prevalent in the USA, can return to normal drinking.

Patterns of drinking are variable and it is a mistake to associate one particular pattern exclusively with 'alcoholism'. America's best-known authority on alcoholism, E. M. Jellinek, divided alcoholics into various 'species' depending on their pattern of drinking. One species, the so-called gamma alcoholic, is common in the USA and conforms to the stereotype of the Alcoholics Anonymous alcoholic. Gamma alcoholics have problems with 'control'; once they begin drinking, they are unable to stop until poor health or depleted financial resources prevent them from continuing. Once the 'bender' is terminated, however, the person is able to abstain from alcohol for varying lengths of time. Jellinek contrasted the gamma alcoholic with a species of alcoholic common in France. The latter has 'control' but is 'unable to abstain'; he *must* drink a given quantity of alcohol every day, although he has no compulsion to exceed this amount. He may not recognize that he has an alcohol problem until, for reasons beyond his control, he has to stop drinking, whereupon he experiences withdrawal symptoms.

A French alcoholic describes himself:

My name is Pierre. I am not an alcoholic. I do not know alcoholics. There are no alcoholics in France, except tourists.

I have drunk wine since I was a child. Wine is good for you. I drink it with meals and when I am thirsty. Since I was a young man, I have drunk three or four litres of wine every day. I also enjoy an occasional apéritif, especially on Sunday mornings and after work. I never drink more than this. I have no problems from alcohol.

Once, when I was in the Army, no wine was permitted. I started shaking all over and thought bugs were crawling on me. I think it was the Army food. My doctor says my liver is too large. My father and grandfather had large livers. It probably means nothing.

The American alcoholic stereotype has two choices—abstain or go on a bender. The French alcoholic stereotype does not go on benders, but cannot abstain.

Although these two types of alcoholism do exist, many individuals who do not conform to the stereotypes still have serious drinking problems. Among American alcoholics, one drink does *not* invariably lead to a binge; a person may drink moderately for a long time before his drinking begins to interfere with his health or social functioning.

Black alcoholics start drinking younger—often in their early and mid-teens. By twenty they may be floridly alcoholic and need hospitalization. They have withdrawal hallucinations more often than do white alcoholics and, for unknown reasons, are less suicidal in middle age.

This diversity in drinking patterns explains the current emphasis on *problems*, rather than a single set of symptoms, as the basis for diagnosing alcoholism.

The most common causes of death in alcoholics are suicide, homicide, accidents, and a variety of medical illnesses, including acute hepatitis, cirrhosis, pancreatitis, subdural haematoma, pneumonia, and alcohol-related heart disease. Apparently this increase in mortality occurs only with the very heavy drinkers we call alcoholic. Studies find the highest life expectancy among moderate drinkers, a somewhat shorter expectancy among abstainers and the shortest of all for heavy drinkers. This does not prove that alcohol prolongs life.* Whatever induces a man to drink moderately may be associated with characteristics that lead to a long life. However, a recent large-scale study found a positive correlation between moderate drinking and longevity, even when other factors (such as smoking, social class) are taken

---

* It is true, though, that man the tippler tends to outlive his cousins of the animal kingdom. Here is a little poem by 'Anon', a favourite of nineteenth century anthologies:

> The horse and mule live 30 years
> And nothing know of wines and beers.
> The goat and sheep at 20 die
> And never taste of Scotch or Rye.
> The cow drinks water by the ton
> And at 18 is mostly done.
> The dog at 15 cashes in
> Without the aid of rum and gin.
> The cat in milk and water soaks
> And then in 12 short years it croaks.
> The modest, sober, bone-dry hen
> Lay eggs for nogs, then dies at 10.
> All animals are strictly dry:
> They sinless live and swiftly die;
> But sinful, ginful, rum-soaked men
> Survive for three score years and ten.
> And some of them, a very few,
> Stay pickled till they're 92.

into account. This observation has gained support from other studies showing that drinkers (non-alcoholics) have less coronary artery disease than do non-drinkers and higher levels of high density lipoproteins (which are associated with reduced risk of heart disease). These findings should be viewed as no more than tentative; moderate drinking is not yet being prescribed to prevent heart disease.

Alcoholism has a higher 'spontaneous' remission rate than is often recognized. The incidence of first admissions to psychiatric hospitals for alcoholism drops markedly in the sixth and seventh decades, as do first arrests for alcohol-related offences. Although the mortality rate among alcoholics is higher than among non-alcoholics, this is probably insufficient to account for the apparent decrease in problem drinkers in middle and late middle life.

Will treatment of alcoholism improve the prognosis? Many believe it does, but mostly on faith. There is little hard evidence one way or the other. Regardless of treatment, most follow-up studies find that about one-third of alcoholics will be abstinent a year after treatment, one-third will be less impaired but still drinking, and another one-third will be unchanged or worse.

Before 1970, most alcoholics in treatment centres could be described as 'pure' alcoholics. If they also used drugs, it was to relieve anxiety and insomnia—often caused by alcohol—or find a substitute for alcohol that was less destructive. Nevertheless, alcohol remained their 'drug of choice'. Since drug abuse became so widespread in subsequent years, more and more alcoholics are also drug abusers, and their 'drug of choice' may be cocaine or opiates rather than alcohol. The polydrug phenomenon is too recent to be fully understood; it has certainly complicated treatment and altered the natural history of uncomplicated alcoholism in ways that are still poorly apprehended.

8 Women and alcohol

For all the effort devoted to eliminating sex discrimination in the work-place, discrimination still exists in science. Most medical studies still deal with men. The same situation applies to studies about alcoholism and alcohol-related problems. Most have been about men, one reason being that men are more often alcoholic than women and it is easy to find male alcoholics if you want to do a study. Matters have improved somewhat in recent years and several books have been devoted to female alcohol problems. For example, after many years of all-male twin studies, a study of women twins was undertaken in the early 1990s. Even in women, heredity apparently plays a major role in alcoholism (see Chapter 11).

Other studies have provided tantalizing distinctions between male and female drinkers. The most talked about distinction arises from the woman's ability to be pregnant. There is impressive evidence that drinking and pregnancy do not mix, as discussed later.

Here are ways in which women alcoholics appear to differ from men alcoholics:

1. Women tend to become alcoholic at an older age. If men are not alcoholic by their mid-forties, they probably will not become alcoholic. This is less true of women.

2. Women alcoholics are more likely to have a depressive illness preceding or coinciding with heavy drinking.

3. Alcoholism in women is more serious. Women are harder to treat and stay sober for briefer intervals. The interval between the onset of heavy drinking and the start of treatment appears to be shorter for women. (Women are more likely than men to seek help for health problems in general, and to do so at an earlier stage, so this 'telescoping' in the development of alcoholism in women may have no connection with the illness itself.)

4. Women alcoholics more often have disruptive early life experiences, such as loss of a parent or other close relative, or psychiatric problems in the family. In several studies about half of women alcoholics reported a parent missing during their childhood, compared with about 15 per cent for male alcoholics.

5. Women alcoholics more often have alcoholism in the family. Many studies have shown a high incidence of alcoholism in a parent or sibling of women alcoholics (upwards of 60 per cent). Women alcoholics are more likely to have relatives who are clinically depressed or commit suicide.

6. Women alcoholics tend to use alcohol medicinally as a form of self-treatment. Male alcoholics tend to use alcohol (at least early on) recreationally and socially. Women alcoholics are more likely to use prescribed as well as over-the-counter medications. They are more likely to leave a doctor's office with a tranquillizer prescription. They are more likely to seek out solutions-in-medications, and apparently doctors encourage this tendency.

7. Women are more likely to cite a traumatic event as the cause of heavy drinking. The event may be a divorce, rejection by a spouse or lover, abandonment, the death of someone close, hysterectomy, miscarriage, or health problem.

8. Women alcoholics tend to be submissive as children, rebellious as adults. This is said to be less true of men alcoholics.

9. Men generally are introduced to heavy drinking by other men when they are young. Women tend to get involved in heavy drinking later in life, often through the influence of husbands or men friends.

10. Women are more likely to have a personality change when drinking; find unexplained bruises after a drinking episode; and drink before a 'new situation' (based on a study of members of Alcoholics Anonymous).

Some of these distinctions are pretty conjectural and probably some do not exist. But all have been reported in respectable scientific journals.

Items to be particularly sceptical about are 6 and 8. Item 6 was obviously based on studies by *men*. Regarding 8, if every submissive child who became rebellious as an adult drank too much, the oceans could not contain enough alcohol to meet the demand. One wonders, too, about the bruises (item 10). Maybe men can't see their bruises because of those hairy legs.

Some more information from the scientific journals follows.

Women have more medical complications—particularly liver disease—from alcohol than men and experience them after fewer years of drinking. They also become drunk on smaller amounts of alcohol. The latter observation was once believed to reflect male chauvinism. (Rarely is a woman ever called a two-fisted drinker, a macho expression of male origin.) If women do indeed get drunk on smaller amounts of alcohol (and certainly this does not apply to all women) it may be because women in general are smaller than men and smaller people get drunk on less alcohol simply because less is required to fill a small space. Two other factors may be important: women in general have more body fat than men and alcohol is less soluble in fat than in water, so the alcohol concentration in blood and other watery parts of the body would be higher as a result of this anatomical difference between the sexes.

Also, as mentioned earlier, recent studies show that women achieve higher levels of blood-alcohol than men after consuming the same amount, because more alcohol in women passes directly into the small intestine where it is absorbed rapidly into the blood stream. The reason, apparently, is that women metabolize less alcohol in the stomach because an enzyme that breaks down alcohol is less active in the stomachs of women than in men.

Then, of course, there is the factor of expectation. Americans have long believed that women cannot 'hold' their liquor as well as men. Many held the view that women should not drink at all. Until recently there were many all-male bars in the USA, and if a woman ventured alone into *any* bar she was automatically classified as a prostitute. Probably the largest single reason for the passage of the Prohibition law in the USA in 1919 was that large numbers of teetotalling women were finally allowed to vote. Presumably, by 1933, when Prohibition vanished, more women drank and enjoyed it and this was reflected in the voting booth.

Attitudes toward drinking by women apparently were different in the UK and the wine- and beer-drinking countries of western Europe. There is the common idea today that women are drinking more than ever, and perhaps this is true in the USA, but apparently not in every country. In 1892 the *British Medical Journal* was 'appalled' by the increase of drunkenness among women in England as well as 'other parts of the world'. The number of women convicted of drunkenness had doubled in ten years, rising from about 5000 to 10 000 in 1884. In Ireland the 'champion female inebriate had been arrested over 700 times, and is yet less than 40 years old'. The *Journal* said that whereas formally men alcoholics outnumbered women alcoholics by seven to one, by 1892 the ratio was three to one. The age of the offenders ranged from 12 to 60. Female alcoholism was called a 'national shame'. Alcoholic women were said to represent all classes and conditions of society. The editor believed this increase of inebriety extended to all civilized countries, but conceded he had no evidence for this. (There were not many statistics in those days.) The *Journal of the American Medical Association* responded to its British counterpart in 1892 by saying there was no problem with female alcoholism in the USA. The Editor went on: 'It may be safely asserted that the American woman cannot constitutionally use any form of alcohol as her foreign sisters use it. She has a more acute nervous organization, and the brain centers are more unstable; the surroundings are full of psychological factors that keep up a certain nerve tension, which antagonizes the sudden increase in the heart's action from alcohol. The brain suffers from the strain of alcohol, which gives no pleasure. The American women of all classes want rest, not increased excitement, hence they seek this more naturally in narcotics.' (There was an epidemic of unregulated opium use by both sexes in the USA and UK in the nineteenth century.) The American editor urged that the British should stop arresting women alcoholics and pleaded that alcoholism was a disease in both sexes and the domain of medicine not the law. 'The time is coming when the medical profession will teach the world the causes and remedies of the great and widespread evil of the century.'

This was written in 1892. One hears the same words today. The 'causes and remedies' have remained elusive.

What *is* the situation today? This is the Age of Statistics and we probably know more about gender differences in drinking than we did a hundred years ago.

It remains true that there are more men alcoholics than women alcoholics. Some believe the gap is closing, but there is no direct evidence for this.* The best current estimate of the sex ratio for alcoholism in the USA is 3:1 (male:female).

With 'liberation' has come an increase in women drinkers. In secondary schools as many girls now drink as boys. This was not true 20 years ago.

But has there been an increase in heavy drinking among women? Probably not. Less than 10 per cent are heavy drinkers, meaning they drink almost every day and get intoxicated perhaps several times a month. Between 20 and 40 per cent of men fall in this category, with young men more likely than older men to be heavy drinkers.

Because more girls are drinking than apparently ever before it is commonly assumed that the rate of heavy drinking and alcoholism will automatically increase in women. This may not occur for one reason:

More women than men are physiologically intolerant of alcohol; after a modest amount of alcohol, more women experience dizziness, headache, nausea, or a sense of simply having

---

* The first British book on the subject, *Women and alcohol*, was published in the autumn of 1980 by The Camberwell Council on Alcoholism. It holds that alcoholism among British women is indeed increasing, based on the following kinds of evidence.

Since 1964 the number of men admitted to psychiatric hospitals in England and Wales with a diagnosis of alcoholism doubled; the corresponding number of women so diagnosed *trebled*. There also was a disproportionate increase in women's rates of convictions for drunkenness (including drunken driving), and for cirrhosis deaths and other forms of alcohol-related mortality. Agencies providing help for problem drinkers recorded a steady increase in women clients.

This may seem persuasive, but increases in treatment cases and arrests do not necessarily mean increases in the population. In recent years, alcoholism has become increasingly destigmatized and more alcoholics of both sexes are 'surfacing' in treatment facilities. Moreover, more treatment facilities exist, and there tends to be a correlation between opportunities for treatment and the number of people who seek treatment. To repeat, there is no direct evidence of an increase in female alcoholism.

enough. Since this 'protection' is physiological and probably genetically determined, presumably it will not disappear even though more secondary school girls drink now and then.

Men and women have different hormones, and hormones in women, if not men, seem to influence drinking behaviour. As mentioned, when drinking the same amount of alcohol, in proportion to body-weight, women have higher blood levels of alcohol than men and these blood levels vary according to the phase of the hormonally controlled menstrual cycle, being highest in the premenstrual period. Many women say they drink more during the premenstrual period and alcoholic women more often give a history of premenstrual tension than do non-alcoholic women. Just how these hormonal shifts relate to alcoholism, if they relate at all, is not known.

The fetal alcohol syndrome

Women have heard for a long time that they shouldn't drink during pregnancy. The idea goes back at least to Biblical times. In Judges 13:7 an angel tells Samson's mother: 'Behold, thou shalt conceive and bear a son: and now drink no wine or strong drink.'

In early Carthage bridal couples were forbidden to drink for fear of producing a defective child. According to Aristotle, 'Foolish, drunken and harebrained women most often bring forth children like unto themselves, morose and languid.' In 1834 a report to the British House of Commons said: 'Infants of alcoholic mothers often have a starved, shrivelled and imperfect look.'

The Biblical injunction does not make clear *why* pregnant women shouldn't drink. The concern may have been that the child would become alcoholic. Plutarch said that 'Drunkards beget drunkards', reflecting a belief held well into the twentieth century that traits and habits acquired by parents would be passed along to offspring. There was also of course the theological view that sins of the fathers, and no doubt mothers, would be passed on.

Only in the last decade has convincing evidence emerged that pregnant women should not drink, or drink much, for another reason: heavy drinking may produce fetal abnormalities. During this decade the term 'fetal alcohol syndrome' came into vogue. It has been widely publicized and has absorbed most of the resources in time and money devoted to research about women and alcohol.

Exactly what is the fetal alcohol syndrome?

It was first described with any precision in an obscure French gynaecological journal in 1968 and again, with much more publicity, in 1972 by two Seattle paediatricians. These reports led to an almost evangelical search for the syndrome in children of drinking women. After many conferences, large amounts of money spent on research, and passionate disagreement, a kind of consensus has emerged.

The fetal alcohol syndrome (FAS) has specific and non-specific features. The specific features largely relate to the head. An FAS baby has a small head. The baby has a short nose, thin upper lip, an indistinct groove between upper lip and nose (called the philtrum), small eye openings, and flat cheeks. These facial features are considered characteristic of FAS, although they are also associated with maternal use of certain drugs (particularly those prescribed for epilepsy).

The non-specific features are multitudinous. They include low birth weight, retarded growth as an infant and child, mental retardation, heart murmurs, birthmarks, hernias, and urinary tract abnormalities.

It now appears that between 30 and 50 per cent of women who drink 'heavily' during pregnancy have infants with one or more of these defects. The incidence of fetal abnormalities in children born of women who drink lightly or not at all during pregnancy appears to be about 5 to 10 per cent. (Abnormality is used here in a broad sense, referring to anything from a birthmark to missing limbs.)

The term 'possible fetal alcohol effects' has been suggested to describe these non-specific abnormalities. It has become increasingly clear that the specific facial characteristics of the FAS are fairly rare.

Some studies have failed to find the FAS in *any* of the offspring of heavy drinking mothers. In one study of 12 000 deliveries it was reported that five of 204 children born of heavy drinking women (2.5 per cent) had the FAS, and even in these few cases the examining doctors knew the mother had been a heavy drinker and may have been deliberately looking for FAS features. ('Seek and ye shall find.') A few studies report lower IQs in children of women who have *anything* to drink during pregnancy, and there is now a warning on alcoholic beverages in the USA telling women to avoid all alcohol. Studies connecting moderate drinking with adverse effects in the offspring are still rare, but perhaps the warning is justified.

One thing, however, seems clear. Women who give a history of 'heavy' drinking during pregnancy have about a fifty-fifty chance of having a baby with some abnormality, and the abnormality most often reported is low birth-weight.

There are two problems in interpreting this finding. One has to do with the definition of 'heavy'. The other refers to the fact that alcoholic women are often heavy smokers and often use other drugs suspected of causing fetal abnormalities; tend to have poor nutrition; fall down a good deal; and generally have a life style different from that of non-alcoholic women.

The definition of 'heavy' remains elusive. It ranges from two or more drinks a day to eight or more, depending on the study. Even heavy drinking women drink erratically during pregnancy, ranging from binge drinking to abstinence and moderate drinking. Timing is crucial in fetal development. In a matter of seconds early in pregnancy a virus or drug can determine whether a limb is formed or not formed; birth size relates mainly to events occurring in the last two or three months of fetal development. It is not surprising that the FAS has come to describe such a welter of observations.

How much a pregnant woman can drink without risk to the fetus is not known and perhaps never will be. Women are simply too diverse, and so are fetuses. To be on the absolute safe side pregnant women should not of course drink at all. Since for the first few weeks after conception many women do not know they are pregnant, this really means women should abstain from alcohol as long as they are able to have babies. In our society this is asking a lot.

There are some sceptics who question whether alcohol alone, in whatever amounts, produces specific fetal abnormalities in the same way that German measles and thalidomide produce specific abnormalities. As mentioned above, alcoholic women differ from non-alcoholic women in many other ways than simply drinking more.

Take smoking. At least 70 per cent of alcoholic women smoke cigarettes. Women who smoke cigarettes tend to have small babies, whether they drink or not, so perhaps cigarette smoking and not alcohol is responsible for the small babies born of alcoholic women.

Because non-smoking alcoholic women are rare, this is a difficult question to answer. However, studies suggest that alcohol abuse and cigarette smoking contribute independently to small size of the newborn. The best estimate at present is that alcohol abuse approximately doubles the risk of birth of a small infant, whether the mother smokes or not.

Another factor to consider is heredity. Birth defects run in families. So does alcoholism. Is it possible that in some instances a common genetic predisposition explains both? There is little evidence for this one way or the other. However, some reports about twins suggest this explanation may not be far fetched in all instances. Sometimes one twin shows signs of the FAS and the other does not; if alcohol is responsible, this should not happen.

There obviously is a great deal of uncertainty about the FAS and how often it occurs. To what extent should women modify their usual drinking patterns because of concern about the FAS? Many women spontaneously reduce their alcohol intake during pregnancy because alcohol makes them ill. Should they reduce their intake to zero?

Some years ago, investigators of the fetal alcohol syndrome made the following recommendation: 'Observations of human babies and of experimentally treated animals have made it clear that a mother's heavy drinking can severely damage her unborn child. We do not know the exact amount or timing of drinking that causes these effects. We cannot say whether there is a safe amount of drinking or whether there is a safe time during pregnancy. We do know that heavy drinking can be damaging. Women should therefore be especially cautious about drinking during pregnancy and when they are likely to be pregnant.' Most authorities still believe this is good advice.

There are thousands of women in Alcoholics Anonymous and in alcoholism treatment facilities who have perfectly normal children and this does not seem to get much attention. Many more children of problem drinking mothers will have to be studied before the fetal alcohol syndrome can be defined with certainty. Meanwhile, caution obviously is the wisest course, both for women who drink and for investigators who make premature and poorly substantiated claims.

The question whether pregnant women should drink raises another: should women drink while breast feeding? There is certainly alcohol in the milk. If the mother is intoxicated while breast-feeding, the infant will be intoxicated.

No one knows whether this will harm the infant. Newborns, like fetuses, are in a plastic stage of development. One would expect them to be especially vulnerable to environmental insults like alcohol. Not drinking for a few hours preceding nursing, and during nursing, is undoubtedly the safest course.

SECTION THREE

Understanding alcoholism

First the man takes a drink, then the drink takes a drink, then the drink takes the man.

—Japanese proverb

9 Risk factors

In 1991, the US Department of Health and Human Services sent a report to Congress called *Alcohol and health*. It identified three types of drinkers:

1. The majority of adult Americans who drink with few, if any, problems.
2. Problem drinkers, also known as 'alcohol abusers', who misuse alcohol because of poor judgment, lack of understanding of the risk involved, or lack of concern about the consequences. These people are not 'dependent' on alcohol in the sense of needing to drink regardless of problems it might cause. They are considered responsible for their behaviour. They often can modify their drinking in response to simple explanations and warnings.
3. Alcohol-dependent drinkers who suffer from a disease called alcoholism or alcoholism-dependence, which, according to the report, has four main features: tolerance; physical dependence; impaired control over alcohol intake on any occasion once drinking has begun; and a discomfort from abstinence that is experienced as 'craving', which can lead to relapse.

Definitions are provided. *Tolerance* means the more a person drinks the more he must drink to attain the same effects as produced previously with smaller amounts. *Physical dependence* means a person experiences withdrawal symptoms when he stops drinking. *Loss of control* is believed by many specialists in the USA to be the *sine qua non* of alcoholism. Not all alcoholics 'lose' control and get drunk every time they take a drink, although some do. The distinctive feature of alcoholism is that the person *cannot be sure* of always being able to stop drinking once he has started. It is this unpredictability, according to Mark Keller, a pioneer in alcohol studies, that best defines alcoholism. Others disagree. Alcoholics, they say, are not like helpless victims of measles or cancer. They may have 'impaired control', but they can gain control through will power or learning certain techniques (discussed in Chapter 15).

In any case, whether alcoholism is something that is learned and therefore can be unlearned, or whether alcoholics are just as helpless in controlling their drinking (without help) as a cancer victim is powerless to shrink a tumour without help, *the fact remains that there is no scientifically acceptable explanation why some people develop problems from alcohol and most do not.* The ultimate goal in alcoholism research is to solve this mystery.

While the cause of alcoholism is unknown, a number of 'risk factors' have been identified. They include the following.

Family history

Alcoholism runs in families. Children of alcoholics become alcoholic about four times more often than children of non-alcoholics. There is evidence that they become alcoholic whether raised by their alcoholic parent or not. (Chapter 11 examines this evidence.) Whether 'hereditary' or not, alcoholism in the family is probably the strongest predictor of alcoholism occurring in particular individuals.

Sex

More men are alcoholic than women. The difference is about 3 to 1, although the ratio varies according to the population studied. In prison populations, male alcoholics outnumber female alcoholics by 12 to 1 taking into account the larger number of men in prison. Throughout history and in every culture studied, men outnumber women in becoming alcoholic. This suggests that cultural factors may not be as important as genetic or physiological factors in explaining the difference between male and female alcoholism rates. Maybe, indeed, hormones play a role, although exactly what role remains unknown.

Age

Alcoholism in men usually develops in the teens, twenties, and thirties. In women it often develops later. People over 65 rarely become alcoholic, regardless of sex. Concern is often expressed about drinking problems in the elderly. Elderly people do sometimes drink too much for people *of their age*. People in their 60s and 70s or older progressively lose tolerance for alcohol and after drinking what younger people would consider rather moderate amounts (two or three beers) develop slurred speech, poor co-ordination, and other evidence of over-imbibing. The important point is that if men are not clearly alcoholic by their mid-40s they are unlikely to become alcoholic later, using the definition for alcoholism in Chapter 5. This does not mean that alcohol may not adversely effect them in later years, and occasionally it does.

Geographical

People living in urban or suburban areas are more often alcoholic than those living on farms or in small towns, at least in the USA. People in northern countries are more often alcoholic than those in southern countries. In France people living in the north are more often alcoholic than those living in the south.

Occupation

Statistics on cirrhosis indicate that individuals in certain occupations may be more vulnerable to alcoholism than those doing other types of work. Waiters, bartenders, dockers, musicians, authors, and reporters have relatively high cirrhosis rates; accountants, postmen, and carpenters have relatively low rates. There appears to be a high incidence of alcoholism in the military services. The group with possibly the highest alcoholism rate in the world may consist of Americans who won the Nobel Prize for Literature. Of seven Nobel Prize winners, five were alcoholic.

Whether particular occupations contribute to the development of alcoholism is not clear. Conceivably bartenders become bartenders because of the availability of alcohol on the job. Some reporters may choose their jobs because the hours and work conditions are favourable to drinking.

Racial background

In the USA, blacks in urban ghettos have a high rate of alcoholism as well as drug use; whether rural blacks have a high rate is unknown. American Indians are said to have a high rate, but this does not apply to all tribes, and the prevalence of alcoholism among Indians generally is unknown.

Orientals have low rates of alcoholism. Many Orientals are physiologically intolerant of alcohol, which may explain this partly.

Income

In the UK the distribution of alcoholism among social classes is believed to be bimodal; the highest rates are in the lower and upper classes, with the middle class having an intermediate rate. It is not clear whether the same is true of alcoholism in the USA.

Money obviously is necessary to buy alcohol, but in recent years alcohol in most western countries has been a bargain compared with most other commodities. The decrease in price of alcoholic beverages in relation to all consumer goods is exemplified by distilled spirits, which increased in retail price by only 22 per cent between 1973 and 1990, while consumer prices generally rose much more rapidly.

(In countries such as Sweden, the price of distilled spirits is tied to the consumer price index so that a consistent relationship is maintained. United States federal taxes on alcohol have changed very little since 1961.)

Whether increasing the price of alcohol results in a decrease of alcoholism is not known. Some countries have tried placing exorbitant taxes on alcohol to reduce alcohol problems, with equivocal results. Gorbachov raised taxes on alcohol in the former USSR, but people just drank more moonshine and there was no decline in overall consumption.

Nationality

The per capita consumption of alcohol in France and Italy is comparable, but alcoholism is less common in Italy. Estimates of alcoholism rates are usually based on cirrhosis figures and admissions to psychiatric hospitals for alcoholism. France may have the highest cirrhosis rate in the world—considerably higher than Italy—but cirrhosis rates are hard to obtain in many countries, including Russia. The USA and UK rank in the middle third of alcohol consumers in the world, based on WHO data. It is interesting that Ireland has a relatively low cirrhosis rate, despite its reputation for rampant alcoholism.

Religion

Almost all Jews and Episcopalians drink, but alcoholism among Jews is uncommon and appears relatively low among Episcopalians. Irish Catholics in the United States and Great Britain have a high rate of alcoholism. Although fewer than 50 per cent of Southern Baptists drink, Baptists and members of other fundamentalist churches in the South appear to have a relatively high rate of alcoholism.

School difficulty

Alcohol problems are correlated with a history of school difficulty. Secondary school dropouts and individuals with a record of frequent truancy and delinquency appear to have a particularly high rate of alcoholism.

These are factors associated with alcoholism. Any ultimate explanation of alcoholism must account for them, at least in a broad way.

10 Alcoholism and depression

Melancholy is at the bottom of everything, just as at the end of all rivers is the sea . . . Can it be otherwise in a world where nothing lasts, where all we have loved or shall love must die? Is death, then, the secret of life? The gloom of an eternal mourning enwraps every serious and thoughtful soul, as night enwraps the universe.
—Henri-Frederick Amiel (1893)

And malt does more than Milton can
To justify God's ways to man.
—A. E. Housman (1894)

God's ways have struck many people, including Amiel, as pretty depressing. What is a good cure for depression? Alcohol, Housman says, and through the ages people have agreed with him.

But is it so? Does alcohol relieve depression? Will it relieve *serious* depression, the kind psychiatrists see?

More importantly, does depression *cause* alcoholism? Are alcoholics really just depressed people who drink to feel better (and better, and better) until drinking becomes uncontrolled, a habit with its own propulsion, progressing independently of the depressed feelings that caused it?

Some believe it. What is the evidence? Here are the questions that need to be asked.

How many alcoholics have manic-depressive disease?*

Many alcoholics become depressed. Their depressions resemble the depressions seen in manic-depressive disease. The alco-

---

* Manic-depressive disease is a condition manifested by episodes of depression which are prolonged, disabling, and often require treatment. Episodes of mania may or may not have occurred. As used here, manic-depressive is synonymous with 'endogenous depression', 'bipolar affective disorder', 'unipolar affective disorder', melancholia, and major depression.

holics become irritable and can't sleep. They feel melancholy and sad. They experience feelings of guilt and remorse. They lose interest in life and contemplate suicide.

Suicide, indeed, is a common outcome of alcoholism. Except for manic-depressives, alcoholics commit suicide more than any other group. One reason to believe alcoholism and manic-depressive disease are related is that, in Western countries, most people who commit suicide have one or the other illness. Many psychoanalysts believe that alcoholism and manic-depressive disease have the same origin; victims of both ill-nesses are orally fixated and, instead of feeling angry toward others, feel angry toward themselves. Aggression directed toward one's self is experienced as a feeling of depression. The ultimate act of self-aggression is suicide. Alcoholism has been called 'slow suicide'.

This is an interesting theory, but difficult to prove scientifically. Although the depression experienced by alcoholics resembles manic-depressive depression, there is a difference: when the alcoholic stops drinking, the depression often goes away. Alcohol is a toxin. In large amounts it produces depression, anxiety, irritability. The alcoholic feels guilty and for good reason: he *has* botched up his life and he knows it.

Based on most studies it appears that alcoholism causes depression more often than depression causes alcoholism. As noted, the cure for alcohol-induced depression is not anti-depressant pills or electroconvulsive therapy but abstinence. (The toxic effects, incidentally, may last for weeks or months after a person stops drinking. Often several months pass before sleep becomes normal again.)

Chapter 8 tells how women alcoholics are different from men alcoholics. One difference is that they more often have depres-sions *preceding* the onset of heavy drinking. In women more than men, a case can be made that alcoholism is sometimes a manifestation of depression. Women, as a group, start drinking heavily at an older age than men. In people who are drinking heavily it is hard to tell whether the depression or alcoholism came first. Since women become alcoholic at an older age than men, there is more opportunity for depressions to occur before heavy drinking. Depression may promote heavy drinking in men as often as it appears to in women, but because men start drinking heavily at an earlier age, it is impossible to determine whether the depressions are a cause or consequence of the drinking.

How many manic-depressives are alcoholic?

This sounds like the same question as the one above, but isn't. If you study a group of manic-depressive patients, how many are alcoholic? Such studies have been conducted, with conflicting results. In one study, one-third of manic-depressives were alcoholic. In another study 8 per cent were alcoholic. Eight per cent is not much higher than the prevalence of manic-depressive disease (as broadly defined here) in the general population. A third is much higher. Which is correct? Nobody knows.

Do manic-depressives drink more when they are depressed?

Leaving aside alcoholic manic-depressives, some drink more and some drink less. Probably more manic-depressives cut down on their drinking when depressed than increase their drinking.

One investigator gave alcohol to hospitalized manic-depressives. A small amount of alcohol improved their mood. A large amount made them more depressed. The same is true of alcoholics. Bring alcoholics into an experimental ward, give them alcohol for a long period, and instead of feeling happy they feel miserable. Alcohol, it turns out, is a rather weak euphoriant compared, for example, with cocaine and amphetamines. The hedonistic explanation for alcoholism, contrary to popular opinion, has little support in science.

After giving alcohol to many psychiatric patients, including manic-depressives and alcoholics, Dr Demmie Mayfield at the University of Texas reached the following conclusions:

If you feel bad, drinking will make you feel a lot better.

If you drink a lot, it will make you feel bad.

Feeling bad from drinking a lot does not seem to make people choose to stop. Feeling a lot better from drinking does not seem to encourage people to continue drinking.

Do manic-depressives drink more when they are manic?

Definitely. Many manics who ordinarily drink little or nothing start drinking heavily when manic. The explanation is not clear but it seems alcohol has almost a specific ameliorative effect on the manic mood. Manics often feel *too* high, uncomfortably high, and alcohol seems to reduce the unpleasant effects of the mania while amplifying the pleasant ones. Lithium, an effective drug for mania, was given to alcoholics and seemed to reduce the frequency of relapse, particularly if the alcoholics were depressed. Other studies—including a large Veterans Administration project—have failed to find lithium useful in alcoholism. The matter is now pretty much closed.

Do alcoholism and manic-depressive disease run in the same families?

If they *did* run in the same families, this would suggest a common genetic predisposition. There is evidence that both alcoholism and manic-depressive disease are influenced by heredity, but is it the *same* heredity?

The evidence is mixed. Some studies show an increase of depression in the families of alcoholics and of alcoholism in the families of manic-depressives. Other studies fail to show this. Those that show an overlap have resulted in a concept called 'depressive spectrum disorder'. In this still hypothetical condition, alcoholism occurs on the male side of families and depression on the female side. There is some evidence, based on adoption studies (Chapter 11), that daughters of alcoholics suffer depressions when they are raised by their alcoholic parents but not when they are separated from their alcoholic parents and raised by non-alcoholic adoptive parents. This suggests that the depression seen in female relatives of alcoholics may be less influenced by heredity than by the environmental circumstance of being raised by an alcoholic parent.

Does antidepressant medication relieve alcoholism?

If it did, of course, this would suggest that alcoholism was caused by depression. The best studies, however, do *not* indicate that antidepressant medication is useful for alcoholism.

To summarize: alcohol may make depressed people feel a little better, but not for long. It may, indeed, make them *more* depressed. The evidence that depression causes alcoholism is weak. Women, more than men, become alcoholic following or during a depression but the causal connection is still not established. Even if depression doesn't cause alcoholism, alcoholism certainly is depressing and most drinking alcoholics are depressed a good deal of the time.

What about other psychiatric illnesses?

In recent years studies have shown a closer association of alcoholism with other psychiatric conditions than depression. There is a strong association with antisocial personality. These are alcoholics, usually male, who are high-risk takers and one of the risks they take is to drink excessively and abuse drugs. They also rob banks and get in trouble with society in a variety of ways. If they survive into their 30s and 40s, people with antisocial personalities (sometimes called sociopaths) often stop their criminal activities and reduce their alcohol intake. This is sometimes called 'burning out'.

Another strong association of psychiatric disorders and alcoholism has been reported. Several studies have linked alcoholism with anxiety disorders. The studies report that about a third of hospitalized alcoholics have a history of an anxiety disorder *before* they become heavy drinkers. Generally, the anxiety disorders are agoraphobia in women and social phobias in men. Agoraphobics are the people who won't leave home because they have phobias and fear they will experience a panic attack if they venture out into a shopping mall or supermarket. Social phobias include fear of giving public speeches or performing socially in a variety of ways, such as eating or playing a musical instrument when others are watching. Social phobias are basically performance fears, a concern that people will judge the phobic person to be inadequate. People with phobias may drink as a self-treatment, in order to feel less anxious; then, as they drink more to feel less anxious, they become dependent on alcohol. However, heavy drinking also *produces* anxiety, particularly during hangovers, and the self-treatment hypothesis remains untested.

To summarize, antisocial personality and phobias may be more related to alcoholism than depression.

11 Heredity

*We are as days and have our parents
for our yesterdays.*
—Samuel Butler

Alcoholism runs in families. This has been known for centuries. Plutarch said that 'Drunkards beget drunkards'; Aristotle said that 'Harebrained and drunken women have harebrained and drunken children'. The nineteenth century medical and religious literature is replete with references to the familial aspect of alcoholism. Clergymen blamed the 'sins of the father' for the transmission of alcoholism from generation to generation. Doctors attributed the transmission to deleterious effects of alcohol on the sperm or egg, a consequence of parental imbibing at the time of conception.

Numerous studies in the twentieth century document the familial nature of alcoholism. About one-quarter of the sons of alcoholics become alcoholic and between 5 and 10 per cent of the daughters. The prevalence of alcoholism in the general population is around 5 per cent in men and 1 per cent or less in women. Having alcoholism 'in the family' increases one's chances of becoming alcoholic by a factor of four or five to one.

Not everything that runs in families is inherited. Speaking French runs in families and is not inherited. It is often hard to separate 'nature' from 'nurture' in conditions that run in families: the same people who provide our genes usually bring us up.

One way to distinguish between the influences of heredity and environment is to compare identical and fraternal twins. Characteristics controlled by heredity should co-exist in identical twins (since they share the same genes) and differ in fraternal twins (whose genes are shared only to the extent that siblings share genes). In other words, assuming that large noses are inherited, both identical twins should have a large nose but one fraternal twin may have a large nose and the other have a small nose. If alcoholism is influenced by heredity, both identical twins would develop the illness more often than would both fraternal twins. Does this happen?

There have been about ten studies of alcoholism in which identical twins were compared with fraternal twins. In general, identical twins more often had similar drinking habits, alcohol-related problems, and alcoholism than did fraternal twins, indicating a genetic influence. However, heredity could not totally explain alcoholism in the twins. In some pairs of identical twins, one twin was alcoholic and the other was not. If genes completely controlled development of alcoholism both identical twins should *always* be alcoholic.

Of course, no one expected this would happen. Environmental factors, such as cultural attitudes and the availability of alcohol, obviously are important. In few inherited disorders is there 100 per cent correspondence between occurrence of the illness in both identical twins.

Another way to determine whether heredity influences the development of a familial disorder is to study people who have been adopted. In the case of non-family adoptions, individuals are raised by different people than those who provided their genes.

The author and colleagues from America and Denmark conducted the first large-scale study of alcoholism in adoptees. It was conducted in Denmark because of easy access to adoption records.

First we studied a group of young Danish men who had a biological parent hospitalized for alcoholism but had little or no contact with the parent because they were adopted away early in life and raised by adoptive parents. We found that 18 per cent of these men were alcoholic at a young age (before 30). The rate of alcoholism in this group was four times greater than was found in a comparison group of adoptees of the same age and sex who did not have an alcoholic biological parent. Except for alcoholism, the adopted-away sons of alcoholics were no more likely to have a psychiatric disturbance than were

adopted-away sons of non-alcoholics. They were no more likely
to be non-alcoholic heavy drinkers. In this study, *just alcoholism*
distinguished the adopted sons of alcoholics from the adopted
sons of non-alcoholics. Alcoholism-related problems in people
who were not alcoholic did not show evidence of a genetic
influence.

We also studied sons of alcoholics who were raised by their
alcoholic parents. They had the same rate of alcoholism as did
the sons of alcoholics who were adopted by non-relatives in
infancy and had no exposure to their alcoholic parent.

In addition, we studied daughters of alcoholics, both those
raised by their alcoholic biological parents and those adopted
away in infancy and raised by non-alcoholic adoptive parents.
Both groups of daughters had high rates of alcoholism compared with the general population. A group of adopted women
without known alcoholism in their biological parents also had
a high rate of alcoholism, raising the possibility that adoption
itself in some way promoted alcoholism in women. As was true
of the men adoptees, the adopted-away daughters of alcoholics,
studied in their mid-thirties, showed a high rate of alcoholism
but no evidence of susceptibility to other psychiatric disorders.

Two other adoption studies subsequently were performed,
one in Sweden and the other in Iowa. Both studies produced
initially the same results as the Danish studies: an increased
prevalence of alcoholism in adopted-away children of alcoholics
with no evidence of an increased prevalence of other disorders.
Later analysis of the Swedish data and expansion of the Iowa
study detected environmental factors influencing alcoholism in
the adoptees (especially heavy drinking by the adoptive
parents) and an association of antisocial personality and alcoholism in male adoptees. The three sets of studies differed widely
in methodology (one was purely a record study), but they
continued to find evidence of genetic factors in alcoholism.

From these studies the concept of 'familial alcoholism' has
developed. This differs from 'non-familial' alcoholism in having
the following features:

1. There is a family history of alcoholism.
2. The alcoholism develops at an early age (usually by the mid-
   twenties).

3. The alcoholism is severe, often requiring treatment.
4. Having alcoholism in the family increases the risk of alcoholism, but not of other psychiatric disorders. As mentioned, there are conflicting reports about the latter point, with some studies finding an increase of antisocial personality and drug abuse in adopted-out children of alcoholics and other studies failing to find this association.

The idea that alcoholism can be subdivided into two types—familial and non-familial—has generated some new research findings. It seems that about half of alcoholics have alcoholism in the family. Of those who have alcoholism in the family, about 90 per cent have *two* or more relatives who are alcoholic. (This is also true of other illnesses influenced by heredity, such as breast cancer and late-onset diabetes.) The younger the alcoholic at time of diagnosis, the more likely it is that there will be alcoholism in the family. Familial alcoholism tends to be particularly severe. Whether any alcoholic can ever return to normal drinking is a subject of controversy, but this possibility seems particularly remote with regard to familial alcoholics.

If alcoholism in some individuals is influenced by heredity, what is inherited? No one knows, but in recent years the US Government has invested large amounts of money in searching for biochemical mechanisms that might explain how alcoholism is transmitted from generation to generation. So far, these studies have produced the following tentative results:

1. Familial alcoholics, when given alcohol, show different brain-wave patterns than non-familial alcoholics. Specifically, there appears to be an increase of alpha-activity brain-waves and also an abnormal brain-wave configuration (technically called P300) that generally reflects attention to stimuli.

2. Changes in neurotransmitter activity are produced by alcohol. Neurotransmitters are chemicals that transmit signals from one brain cell to other brain cells. Changes in four neurotransmitters have been found. These are serotonin, GABA, opiate-like neurotransmitters called endorphins, and an adrenaline-like neurotransmitter called norepinephrine. The strongest evidence supporting the role of one of these chemicals in moderating drinking activity consists of studies showing that

serotonin brain levels are increased when drinking and decreased after drinking (permitting a situation that parallels the
'addictive cycle' described in Chapter 12). Other evidence supporting serotonin as a determinant of drinking behaviour is the
fact that drugs that enhance the activity of serotonin (such as
Prozac) at least temporarily reduce drinking in heavy-drinking
people and also consumption of alcohol by animals. Other
studies have shown that serotonin activity is low in familial
alcoholics. At the moment, serotonin is the hottest item (in this
author's opinion) in the search for a biochemical explanation for
what appears to be a genetically determined transmission of
alcoholism.

The ultimate jackpot in alcoholism research would be identi-
fication of a single gene or group of genes that influences drinking behaviour. One such gene, involving the neurotransmitter
called dopamine, was found associated with alcoholism in
several studies but not in several others. The importance of this
gene remains to be determined. Other genes that regulate
serotonin metabolism are being studied. It is often said that
alcoholism is a heterogeneous condition and therefore must be
influenced by many genes, but the evidence for this remains
speculative. Indeed, there is some evidence that the most severe
form of alcohol-dependence called alcoholism is *not* on a continuum with heavy drinking but is a separate disorder, just as
malignant hypertension is probably not on a continuum with
high blood pressure (there are also controversies about high
blood pressure). When an 'alcogene' has finally been identified,
if ever, it may turn out that a single gene determines whether
a person is alcoholic or non-alcoholic (assuming the right set of
environmental factors are present, such as the availability of
alcohol). It is less likely that a single gene will be identified with
what we call alcohol-related problems that less severely damage
the individual.

Geneticists are also searching for genes that might explain
why a small minority of alcoholics develop brain damage and
liver disease and most do not. This is one of the great mysteries
in the field of alcohol studies. There is some evidence that
cirrhosis indeed runs in families, but there is also conflicting

evidence. In any case, discovering why some individuals develop these devastating conditions and most do not would be a medical advance of great importance and probably lead to better treatments and even cures.

There is one form of alcohol-related problem that is definitely inherited and it is called the 'Oriental flush'.

Millions of people have unpleasant reactions to small amounts of alcohol. These may take the form of dizziness, nausea, or headaches. Adverse reactions to alcohol have been most studied in Orientals. About two-thirds of Orientals develop a flush of the skin and have palpitations and other unpleasant effects after drinking a small amount of alcohol. Oriental babies given small amounts of alcohol also develop a flush. The basis for the flush unquestionably is genetic. Apparently this response can be blocked by antihistamine drugs, suggesting the phenomenon may be allergic in nature.

There is some evidence that a higher proportion of women than men have adverse reactions to small amounts of alcohol. There is a lower alcoholism rate in the Orient than in western countries, and this is usually attributed to culture. Women have a lower rate of alcoholism than men and this also has been ascribed to culture. Culture may indeed be important in both instances, but obviously in Orientals, and perhaps in women, another factor contributing to the low alcoholism rate may be physiological and presumably genetic in nature.

Many people, in short, are born *protected* against becoming alcoholic. Conversely, those who become alcoholic are born *un*protected.

Is there more to the story than this? Are children of alcoholics vulnerable to alcoholism only because they lack an inborn intolerance for alcohol?

The story clearly is more complicated. People differ in their response to alcohol in other ways than having varying degrees of intolerance for alcohol. Some people get higher on alcohol than others. As will be explained in the next chapter, people who get higher may also get lower and this rollercoaster effect may explain why some people are more susceptible to alcoholism than others.

12 *The addictive cycle*

Any ultimate explanation of alcoholism must account for two features of the illness: *loss of control* and *relapse*.

Loss of control refers to the alcoholic's inability to stop drinking once he starts. Relapse is the return to heavy drinking after a period of sobriety, and it is the great mystery of addictions. Why, after months or years of abstinence, does the smoker smoke again, the junkie shoot up again, the alcoholic fall off the wagon?

Here is an attempt to show how inherited and environmental factors may combine to produce loss of control and relapse.

Some people experience more pleasure, or glow, from alcohol than others. How much pleasure a person gets from alcohol may be partly determined by heredity. The pleasure is short-lived and, in some people more than others, it is followed by a feeling of discomfort. The *degree* of discomfort may also be determined by heredity (Fig. 12.1).

Alcoholics learn that the discomfort has a simple remedy: another drink. Thus, the alcoholic drinks for two reasons: to achieve the pleasure and to relieve the discomfort. The same substance that produces the happy feeling also produces the unhappy feeling and is required both to restore the one and abolish the other. Nothing abolishes the unhappy feeling quite as effectively as the drug that produced it.

The unhappy feeling is called craving. To relieve craving the alcoholic will try anything—a chocolate bar, sex, tranquillizer, jogging, or prayer—but has learned that only alcohol gives complete and immediate relief from craving. After a time, the alcoholic drinks more to overcome the unpleasant effects of alcohol than to attain the pleasant effects.

Thus some people are 'born' to have higher highs and lower lows from alcohol than others. The highs and lows (the addictive cycle) may occur repeatedly in a single drinking session, and of course there is the monumental low that comes the next day, known as hangover. The mind and body of the alcoholic 'learns' that the lows can be banished by another drink. The learning is experienced physically as craving and psychologically as a preoccupation (literally an obsession) with having alcohol on hand at all times. Once the true binge drinker has started drinking, he

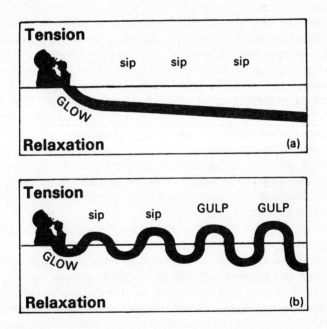

Fig. 12.1 Two different reactions to alcohol. According to the 'addictive cycle' theory, the normal drinker experiences a progressive glow, or relaxation, from alcohol (a). The alcoholic's glow is temporary, followed by a feeling of tension, or craving, that requires another drink for relief (b). These different responses to alcohol may be determined by heredity. (Reprinted with permission of the Schnick Institute.)

often cannot stop until the high–low cycle has left him exhausted and he must stop.

But why start in the first place? Why does the binge drinker, who actually has come to hate alcohol and doesn't drink anything for months or years, one day start drinking again? It is not from ignorance; he knows what will happen. Why go through it *again*?

This brings up the problem of what is and isn't voluntary, which is a philosophical issue having to do with free will, and not a suitable topic for discussion here. Leaving aside free will, relapse can partly be explained by something called *stimulus generalization*.

The term refers to the fact that things remind people of things. For the drinker, the hands of a watch pointing to 5 p.m. (the stimulus) may remind him that, for years, he always had a drink at 5 p.m. (the generalization), and so he drinks. Acts of drinking become embedded in a maze of reminders. Every drinker has his own reminders, but there are common themes. Food, sex, holidays, football games, fishing, travel: all have nothing intrinsically to do with drinking but all commonly become *associated* with drinking and are powerful reminders. Physical feelings become reminders: hunger, fatigue. Moods become reminders: nostalgia, sadness, elation. Anything, in short, can be a reminder and *remain* a reminder long after a person has stopped drinking.

Reminders may lead to relapse. One day, unexpectedly, the 'recovered' alcoholic is flooded with reminders. It is 5 p.m. on Christmas Eve (which also is his birthday). The boss chewed him out and he missed lunch. His alimony cheque to his wife bounced, but he learns he just won the Irish Sweepstakes. Suddenly he has an incredible thirst. As he passes a pub, a strong west wind blows him in the door—and a relapse occurs.

This is an extreme example. The relapse trigger may be subtle.
'For any alcoholic', Mark Keller writes, 'there may be several or a whole battery of critical cues or signals. By the rule of generalization, any critical cue can spread like the tentacles of a vine over a whole range of analogs, and this may account for the growing frequency of bouts, or for the development of a pattern

of continuous inebriation. An exaggerated example is the man who goes out and gets drunk every time his mother-in-law gives him a certain wall-eyed look. After a while he has to get drunk whenever any woman gives him that look.'

In either case, he probably will not be able later to say why he started drinking again. And maybe stimulus generalization is not the whole story. But it seems to explain a lot.

The idea of addictive cycles has been applied not only to alcoholism, but also to thrill-seeking, over-eating, and love. The theory holds that every 'addiction' eventually produces its opposite; pleasure turns to pain, and pain to pleasure.

Proponents of this idea believe that every event in life that has a strong effect also has an opposed process that fights it. 'At the start', one says, 'drugs are highly pleasant. You get a big "rush" and euphoria. But as tolerance builds up, the rush disappears and the threat and pain of withdrawal begin to take command.' The addictive cycle from drugs is similar to a runner's 'high', an example of pain giving way to pleasure. Parachute jumpers sometimes become extremely distressed when bad weather cancels their sport, reflecting, some believe, an 'addiction' to jumping.

Some animal studies seem to support the idea of addictive cycles. In experiments measuring the distress calls of ducklings, birds show far more distress when their mother is removed and returned at brief intervals than when removed for long periods of time. Frequent separations, in short, produce an addictive cycle in which the distress calls are the equivalent of withdrawal symptoms.

Such studies suggest that in its early stages any attachment is controlled mainly by pleasure, but late in the attachment the main control is the threat of separation and loneliness. Although the leap to human behaviour is a long one, some see the same mechanisms at work. 'The ecstasy and madness of the early love affair are going to disappear', the psychologist Richard Solomon observes, 'and when they do, it means that a withdrawal symptom has to emerge if you are denied the presence of your partner.'

According to the theory, the size of doses and the intervals between doses are crucial to addiction. The distress shrieks of ducks are prolonged at one-minute intervals away from the mother, but not at two or five minutes. A rat fed a food pellet every 60 seconds shows withdrawal symptoms (agitated behaviour, drinking too much water) after each morsel. But the symptoms disappear if the pellets are spaced several minutes apart. The implication: proper timing of dosage prevents addiction.

Why do people keep eating when their stomachs are full? 'Because we like to fight off withdrawal by redosing with a pleasurable taste', says Solomon. 'The better the taste of the food, the harder the withdrawal.' It makes sense, he says, to eat tasty foods early in the meal and save bland ones for last, so the withdrawal will be easier. Better yet, eat only bland, uninteresting foods.

A weakness of the theory is that it seems inadequately to take into account the *strength* of addictive behaviour. It is hard to believe that fat people would become thin merely by eating bland food at the end of meals. Still, who knows? People dependent on alcohol or other 'recreational' drugs respond to a description of the addictive cycle by saying, 'Yes, that's the way it is'.

13 *Psychosocial theories*

The fish is in the water and the water is in the fish.
—Arthur Miller

A tendency to alcoholism may be inherited, but most authorities believe alcoholism also is partly learned (and thus psychological) and the learning takes place in a particular social environment—hence the term 'psychosocial'. Which is most important—heredity or learning—is not resolved, and the mix may differ in different people, but the issue is not purely academic. Treatment often depends on the treater's judgment about whether biological or psychosocial influences predominate.

Therapists who believe that alcoholism is mostly learned tend to believe that it can be unlearned and moderate drinking can be a proper goal of treatment. Those who minimize the role of learning tend to view alcoholism as an incurable condition for which total abstinence from alcohol is the only appropriate goal. The former group largely consists of psychologists; the latter mainly consists of physicians and members of Alcoholics Anonymous. Treatment differs in different countries depending on which approach is popular. In the USA, where attitudes toward alcoholism are strongly influenced by Alcoholics Anonymous, the total abstinence philosophy prevails; in the UK, where Alcoholics Anonymous is less influential, it is widely held that many alcoholics—particularly moderate alcoholics—can 'return' to moderate drinking.

Sometimes the term *biopsychosocial* is substituted for psychosocial, meaning the therapist acknowledges that biological factors are important while continuing to maintain that psychological and social factors are at least equally important. This point of view has a certain face validity and is a politically 'correct' view in that it pleases the biologists as well as psychologists and sociologists, all toilers in the vineyards of alcoholism research and treatment. But presumably there is more to the 'multivariant' point of view than diplomacy. Each component of the biopsychosocial perspective is rooted in observation.

Alcoholism, for example, has been called a bad habit and certainly it is, if bad habit means conditioned reflex. Chapter 12 talks about the importance of conditioning—semi-involuntary responses to stimuli, or cues, that trigger a need to drink. Conditioning is in the domain of psychology, even if mediated in the brain. But if alcoholism is a bad habit, it is a bad habit some people acquire easier than others. It seems clear that people are 'born' with a susceptibility to certain bad habits and a resistance to others. Heredity seems to explain this variability more than any other influence.

As for psychosocial factors, people presumably drink for reasons, and the reasons may be psychological (a wish to feel happy) and have social origins (getting drunk at a fraternity beer bust because everyone does it). But psychological theorists do not rely merely on such commonplace examples. The *learning model*, as it is called, gains its strength to some extent from studies where alcoholic beverages are given to people and they react as if they were drunk even if the experimenter cheats and doesn't put alcohol in the drinks. The more alcohol-dependent the subjects are the more they *expect* alcohol will cause them to behave in a particular way, and sure enough it does, for the reason that they have been conditioned by long experience to react to drinking cues and they react that way even when the vodka and tonic contains nothing but tonic. They *learned*, in other words, to 'lose control' (believed by some to be the *sine qua non* of the 'disease model' of alcoholism) and they unlearn it with proper treatment. So the theory goes. Nobody denies the role of expectation in drinking behaviour (see Chapter 3). But few believe that expectation alone causes a person to become alcoholic. Some do, however, and the explanation often goes as follows: the child of an alcoholic says to himself, 'Dad was an alcoholic and therefore I'll become one', and he becomes one.

Not even the most ardent learning theorist believes it is that simple. A shot of whisky, downed straight, has effects that far exceed those based solely on expectation. You can fool an experimental subject when a small amount of alcohol is diluted in tonic water, but no one is fooled when a large amount of alcohol is added. There are biological responses to alcohol, if consumed in sufficient amounts, that cannot be explained by expectation or learning.

There also is the *social model* that emphasizes the importance of availability of alcohol, peer pressure, and cultural values, attitudes and mores that regulate the use of alcohol. Alcoholism, in this model, is viewed as a 'social career' more or less like becoming an accountant or mob hitman. Like the learning model, the importance of social factors cannot be denied. First, there must be alcohol for people to drink alcohol, and society usually provides it. In some Islamic countries possession of alcohol is a serious crime, even punishable by death, and thus is functionally unavailable even if the ingredients for making alcohol— yeast, sugar and water—are ubiquitous. Making alcohol very expensive is another possible way to limit its availability. This was tried in Russia recently and didn't work. There was simply more home-brew made in people's garages. But studies show that in other countries where citizens are less passionately dedicated to drinking, increasing the price of alcohol does indeed reduce its consumption, at least among those who do not consume much anyway. As was shown in Russia, alcohol can be difficult to obtain but easy to make. There is no evidence that higher prices reduce drinking among the dependent drinkers we call alcoholic. Nor do other social measures, such as closing down bars early or taking drastic action against drinking drivers, seem to reduce drinking by the heaviest drinkers. Alcoholism rates vary from country to country and in many Western countries these rates are probably lower today than they were a hundred years ago. However, pricing and accessibility do not appear to be the determining influences. Consider the popularity of the mini bar in hotel rooms. Accessible 24 hours a day for the occupant (and his children!), the mini bar has flourished despite an overall reduction in consumption of alcohol.

Social pressure has been the main factor in reducing cigarette-smoking in Americans (from 50 per cent of adults to 25 per cent) and *not* warning labels on the packages. Social pressure, more

than anything else, has probably contributed to the more
modest decline in consumption of alcohol. Just as non-smokers
finally decided that smoking was bad not only for smokers but
for others around them, alcohol consumption declined when
the majority of drinkers decided that a drunken minority was
not only a nuisance but dangerous on the highway. This, at
least, is one explanation for the alcohol decline, although, again,
it is hard to know whether there are fewer alcoholics.

It has been said that the three basic needs of mankind—food,
security, and love—are so intertwined that it is difficult to tell
whether a person is searching for one or the other. Thus, is it
hard to separate biological, psychological, and social factors in
alcoholism. The more confusing an issue is, the more people
will discuss it, and the following discussion illustrates the point.

A debate*

> *To drink is a Christian diversion*
> *Unknown to the Greek and the Persian*
> —William Congreve

Any final explanation for alcoholism must explain more than
why alcoholism runs in families. It must explain the following:

Why is there a low rate of alcoholism among Jews and a high
rate among the Irish?

Why are there more French alcoholics than Italian? Why are
there more alcoholics in northern France than in southern
France?

Why is alcoholism rare in the Orient?

Why is alcoholism more common in cities than in rural areas?

Why are more reporters alcoholic than mail carriers? More
bartenders than bishops?

Why are more men alcoholics than women?

Imagine a hereditist and an environmentalist (perhaps over a
drink) discussing these statements point by point. Their argument might be as follows.

* Parts of the 'debate' appeared in *Is alcoholism hereditary?* Random House,
1988. It has been revised to accommodate new information.

H: Jews intermarry. Hispanic Jews, for example, are unusually susceptible to a blood disease known to be inherited. Why is it not reasonable to suppose that Jews would be 'protected' from developing alcoholism by a similar hereditary factor?

E: Hispanic Jews do intermarry, but this is much less true of Jews in other places. In America, for example, Jews frequently marry non-Jews, and it is inaccurate to speak of a genetically pure Jewish 'race'. Nevertheless, despite intermarriage with other ethnic and racial groups and increasing assimilation of Jews into the general American culture, their alcoholism rate remains low. If heredity cannot account for this, it must be attributed to environmental factors. Even with assimilation, Jewish tradition continues to influence the way Jewish mothers and fathers rear their children. Traditionally, moderate drinking is condoned, but drunkenness, particularly chronic drunkenness, is condemned and viewed not only as an individual weakness but a family disgrace. In families where some Jewish tradition persists, closer family ties prevail than generally exist among non-Jews. These factors provide a more plausible explanation for the low alcoholism rate among Jews than does any genetic explanation.

E (*continuing*): As for the Irish, no one ever accused them of being a race. A very high proportion of Irish have emigrated to other countries where intermarriage with people of other ethnic and racial backgrounds has been widespread. Again, family and cultural traditions provide a more plausible explanation for high rates of alcoholism among the Irish than heredity does.

E (*continuing*): As for the French and Italians, it is true that people from the two countries differ to some extent in physical characteristics, but there is no known instance of people living in France being more prone to a particular genetic disorder than people living in Italy. Again, there is frequent intermarriage with people from other countries. If indeed there are more alcoholics in northern France than in southern France, this would further support the importance of environmental factors. Perhaps climate is a factor. Since the north is chillier than the south, perhaps northerners drink more to keep warm. It sounds far-fetched but no more so than a genetic explanation, considering that northerners and southerners regularly intermarry.

H: Three points need to be made. In the first place, are the French really more often alcoholic than Italians, and does alcoholism really occur more often in northern France than in southern France? True, this has been reported, but estimates of the prevalence of alcoholism are notoriously unreliable. They are usually derived from cirrhosis figures or admissions to hospitals for alcohol problems. Cirrhosis figures themselves are not very reliable, and whether a person becomes 'officially' an alcoholic by receiving treatment for the condition is influenced by many factors other than the presence of alcoholism itself. One is whether facilities are available for treatment. Another is the readiness of individuals to seek treatment. Another is the diligence with which public officials maintain treatment records. Perhaps in northern France the authorities keep better records than in southern France. This could entirely explain the alleged difference.

At any rate, some people have challenged the repeated assertion that alcoholism is more prevalent in France than in Italy, as indeed some have questioned whether alcoholism is all that common in Ireland. Everyone agrees that Irish immigrants in the USA seem particularly susceptible to alcoholism. Ireland itself, however, has a fairly low reported rate of alcohol-related liver disease, and some people question whether alcoholism is any more common in Ireland than, say, in England. Perhaps in rural areas it may be less common, if only because rural Ireland is less affluent and people have less money to buy alcohol.

This brings us to the second point: alcoholism can only occur when alcohol is available. It is often said that prohibition in the USA was a failure. In some ways, it was a huge success. Alcohol-related liver-disease rates dropped precipitately, as did admissions to hospital for alcohol problems. Similarly, in France during World War II, when wine was rationed, alcohol-related liver disease became less common. In Scandinavia there has apparently been some success in reducing alcohol consumption by increasing the cost of distilled spirits. Perhaps, if it is true that Italy has less alcoholism than France, it is partly because Italy is a poorer country and fewer people can afford distilled spirits.

Finally, even if these differences between cultures and countries are genuine, what we call alcoholism may be a mixed bag of conditions. Some kinds of alcoholism may be genetically influenced while others are not. Because of varying definitions of alcoholism and unreliable means of estimating the prevalence of alcoholism, it is difficult to say what proportion of alcoholics have what might be called 'familial alcoholism', as opposed to alcoholics whose excessive drinking has other causes.

H (*continuing*): As for the rarity of alcoholism in Oriental countries, again one might question the accuracy of the observation and also raise the possibility that officials in such countries as Communist China might try to conceal whatever alcoholism exists so as to make their system of government look good (as, indeed, once occurred in Russia). Assuming, though, that alcoholism is rare in most Oriental countries (and the assumption seems more justified than most) here we have an instance where there is a known physiological difference between one race and other races in their response to alcohol. Several studies have indicated that many Orientals have a low tolerance for alcohol. After drinking a small amount, they flush, develop hives, feel nauseated, and generally are discouraged by their physical reaction to alcohol from drinking more. It is unlikely that these responses are produced by psychological factors arising from social pressure; there are fewer taboos against drinking in the Orient than there are in the American Bible Belt. The low rate of alcoholism in the Orient, in other words, may really reflect a biologically determined low tolerance for alcohol.

E: There are also many non-Orientals who have a low tolerance for alcohol. Relatively few may develop skin flushing, but many people, after a few drinks, become dizzy, develop a headache, or feel sick. Thus large numbers of non-Orientals also are deterred by their physical response to alcohol from drinking excessively. Nevertheless, this leaves considerable numbers of people without low tolerance and from this group America alone produces several million alcoholics. The proportion of Orientals who cannot drink much because of adverse effects from alcohol may be higher, but many can drink as much as they please, and since alcohol is ubiquitous (all that is required is yeast, sugar, and water), one would expect higher rates of alcohol problems in the Orient than apparently exist, unless social factors discourage alcoholism.

E (*continuing*): There is another point about the importance of culture and tradition. Anthropologists have observed that every society tends to have its drug of choice, its 'domesticated intoxicant' that is favoured by the great majority of people over other intoxicating substances. In China, for example, the favoured intoxicant for centuries was opium. In India, the Middle East, and North Africa cannabis derivatives (hashish, marijuana) historically have been the most widely used intoxicant. In Judeo-Christian societies alcohol has been the drug of choice.

In recent years, marijuana has become a competitor to alcohol among the younger members of Western countries. This cultural heresy, as it were, originally was met with strong legal and moral resistance, which has abated considerably in the past ten years; nevertheless, whether marijuana in the West is a passing fad or will remain popular is uncertain. In the past, attempts by other societies to have more than one 'approved' intoxicant have ultimately failed. An example is the religious taboo against alcohol among Moslems, a taboo coupled with religious and legal sanctions as stern as the long prison sentences once meted out to marijuana users in the USA.

Now it is very difficult to explain these cultural differences in intoxicant choice to inherited factors. It is more logical to attribute them to historical accident that led to the selection of one substance and to the development of customs that control its use with minimal damage to society. The 'accident' may have been nothing more than the availability of the substances, such as cannabis in India, the poppy in China, and cocaine from the coca leaf in the Peruvian Andes.

H: All this is true, but it does not explain why certain users of these substances in every society get into serious trouble from their use while apparently the majority use the substance with little or no ill effects. This raises an important scientific question: Are some individuals born (if you'll pardon the expression) with a non-specific vulnerability to addictiveness—with the substance to which they become addicted determined by availability and cultural sanction—or is addiction-proneness specific for certain substances? In other words, is it possible that some individuals have a specific vulnerability to alcohol abuse but can use opium products, nicotine, and other addictive substances within moderate bounds?

The question is still open. Favouring a non-specific addiction proneness is the fact that most alcoholics are heavy smokers and that heroin addicts, if denied heroin, often become alcohol abusers. It would be interesting to know whether opium and cannabis abuse runs in families, as alcohol abuse does. At the moment there is little information on the subject.

Nevertheless, while it is true that there are cultural variations in the preference for intoxicating agents, it is also true that usually only a minority of individuals exposed to the agent develop serious problems from its use, and that just as the hereditist must explain cultural differences in alcohol use, environmentalists must explain why only a minority of users become abusers.

H (*continuing*): This brings us to the next question: Why is alcoholism more common in cities than in rural areas, more common in some occupations than in others, and more common in men than in women?

No one questions the importance of custom and tradition in determining how many people use a particular intoxicating substance such as alcohol. In the USA, for example, there once were social sanctions against women drinking at all, much less becoming intoxicated. There have been religious sanctions against any drinking at all in rural areas where fundamentalism is strong. If a person does not drink, he or she cannot become alcoholic. As society relaxes its pressure concerning the use of alcohol by women (with women freely permitted to go to bars alone and so forth), it is quite likely that the number of alcohol-abusers among women will increase (although, as noted earlier, it is hard to prove).

There is also some evidence that the proportion of abusers to users remains constant. In other words, if the abuse potential for alcohol is, for example, one in ten, 10 per cent of men and women drinkers will be alcoholic. As the number of women drinkers increases, so also will the number of women alcoholics. This, again, indicates that factors other than social and cultural ones (which no doubt contribute to whether a person drinks or not) are operative in selecting those who ultimately will drink too much. The 'selector' may well be a biological, genetically driven susceptibility.

H (*continuing*): As for occupational differences in alcoholism rates, they can be explained fairly easily. Some occupations are more tolerant than others of drinking and heavy drinking. For example, by the nature of his work, a bartender or house painter has more freedom to drink than does a chest surgeon or an airline pilot (both of whom risk censure or worse just by having alcohol on their breath). People inclined to drink heavily very likely gravitate to occupations where heavy drinking is tolerated —if not actively encouraged (as was the case in public relations, until recently at least, with its tradition of the two- or three-martini lunch). People rarely go into various jobs purely by chance. Of the many factors that influence job choice, one, for some, may be the opportunity it provides to drink. This is an instance where tradition and opportunity open the door for genetic traits to walk in. It also shows how group differences in alcohol use, ostensibly resulting from nurture, may actually be nature having her way incognito.

E: William James said something like the following: Toss down a handful of beads and, by ignoring certain beads, you can perceive any pattern you wish. In looking for a genetic explanation for cultural variation in drinking patterns, it seems that is precisely what you are doing. In your review on alcoholism and heredity (I assume that chapter was yours), a progenetic bias also seems obvious. Let me quickly summarize the studies presented in Chapter 11.

The adoption studies are fairly consistent. Sons of alcoholics often become alcoholic even when raised by adoptive parents. It is less clear whether this occurs also in daughters. And there are two adoption studies that do not show an association between parental alcoholism and drinking by adoptees in either sex.

Devoted hereditists either ignore these studies or explain away their findings.

As for twin studies, only three really address alcoholism. Two found evidence favourable for a genetic hypothesis. The other did not. Which should one believe? Granted, the studies defined alcoholism somewhat differently, but this would hardly seem to explain the discrepancy. Most of the other twin studies relied on questionnaires that included a few items about drinking buried in a mass of questions about other matters. One study used telephone interviews! Do you really think anyone with a drinking problem would confess this to a stranger by telephone? It's hard enough to get people to admit to a drinking problem face to face, or tell their own trusted doctor. Questionnaires and telephone calls are absurdly inadequate methods for obtaining such sensitive information. Nevertheless, when the information obtained fits the hereditist's bias, it is usually accepted as incontrovertible truth.

As for animal studies, success in a few labs in creating 'alcoholic' rats was interesting. However, they were produced in a way that fits no conceivable human circumstance: inbreeding for a single variable over many generations. There is something called assortative mating in humans (a tendency for people to marry people of similar interests and backgrounds) but this is not even remotely comparable to multiple generation inbreeding of a type most commonly identified with horticulture.

Moreover, drinking behaviour by rodents is not entirely controlled by genes. This is demonstrated by at least two studies. Both involved two strains of mice. One strain spontaneously drank more alcohol than the other. Everyone assumed this was because of genes. Then the investigators did an 'adoption' study. They took the babies of the high-preference strain and 'adopted' them out to female mice of the low-preference strain. They also did the reverse, having newborn mice of the low-preference strain raised by females of the high-preference strain.

The results were interesting. The baby mice from the high-preference strain reared by mothers of the low-preference strain spurned alcohol; baby mice of the low-preference strain reared by mothers of the high-preference strain had an increased preference for alcohol.

How can this be explained? The answer is that even baby rats learn from their mothers.

E (*continuing*): Let me summarize what I believe are the strongest arguments for both sides of the issue. The strongest evidence for heredity is that alcoholism runs in families, even when the children are separated from the alcoholic parents and raised by adoptive parents. The strongest evidence that social factors contribute to alcoholism is the great diversity in alcohol-use and alcohol-abuse among various cultures, nations, ethnic groups, social classes, regions, sexes, and other groupings. With enough ingenuity, the hereditists can see heredity rearing its head even in these domains, but, for the scholar without a progenetic bias, attributing these differences to heredity seems far-fetched.

The crucial question

As our two polemicists finish their argument and their drinks, pay the bill, and depart into the night, let me repeat a question already raised by the hereditist and deal with it as it might be dealt with by the environmentalist. It is the crucial question. In the USA most adults drink (roughly 70 per cent) and one out of every twelve or fifteen becomes alcoholic. Why are children of alcoholics more vulnerable to alcoholism than children of non-alcoholics?

Here is some non-genetic conjecturing.

1. *Alcoholics make bad parents.* H. J. Clineball, a minister who has counselled many alcoholics and their families, lists four ways in which they make bad parents. First, when one parent is alcoholic, there may be a shift reversal in the parent's roles, complicating the task of achieving a strong sense of sexual identity in the children. Second, an inconsistent, unpredictable relationship with the alcoholic parent is emotionally depriving.

Third, the non-alcoholic parent is disturbed and therefore inadequate in the parental role. And fourth, the family's increased social isolation interferes with peer relationships and with emotional support from the family.

The big problem with the bad-parent explanation is that there are far more bad parents than there are alcoholics. Many children are raised by only one parent or by no parents. With our current high divorce rate, broken homes have become increasingly common. It is unlikely that alcoholism rates have increased proportionately.

Proponents of the bad-parent theory usually deal with these arguments by saying that alcoholism springs from an upbringing that was bad in a specific way. However, there is no agreement about what the specific way is.

2. *Children learn from their parents.* This explanation works in two ways. Observing the havoc produced by their alcoholic parent (or parents), children resolve never to touch alcohol and become adamant teetotallers. Undoubtedly this occurs, but apparently not very often.

The reverse is that children model themselves after their parents; their father or mother (or both) drinks and therefore they drink.

'Dad was only really happy when he drank,' observed the son of an alcoholic. 'Sure, he was sometimes mean when he drank and often an embarrassment to the family. Nevertheless, when he was sober he always seemed tense, withdrawn, unhappy. I much preferred to see him drink, if it wasn't too much. He became jovial, interested in the children, accessible. I came to identify good feelings and a warm and sociable manner with drinking.'

Parents also transmit tradition. If they speak English, their children speak English. Social attitudes toward drinking pass from parents to children, who acquire them as naturally as they do their parents' accents and table manners.

Social values, including those regarding alcohol, find expression in some families more so than in others. A good example is the family of Eugene O'Neill, the playwright. When O'Neill

began drinking, according to biographers, he was simply following the example of his father and brother. His father drank
daily, usually in bar-rooms. 'You brought him up to be a
boozer,' says Eugene's mother to his father in *Long Day's Journey
into Night*. 'Since he first opened his eyes, he saw you drinking.
Always a bottle on the bureau in the cheap hotel rooms!'

To some extent, O'Neill's father was a conduit for Irish
attitudes toward alcohol prevailing at the turn of the century.
When Eugene was an infant and had a nightmare or stomach-
ache, his father gave him a few drops of whisky in water.
O'Neill later believed this old Irish custom contributed to his
drinking problem as an adult.

'I'm all Irish', O'Neill said, referring not only to his ancestry
but also to the Irish customs and attitudes of his family, exquis-
itely portrayed in *Long Day's Journey into Night*. In the play, as
John Henry Raleigh pointed out, a bottle of whisky is at the
centre of the room—in many ways its most important object. 'If
not using it they talk about it. It enters into their very character;
the father's penuriousness is most neatly summed up by the fact
that he keeps his liquor under lock and key and has an eagle eye
for the exact level of the whiskey in the bottle . . . by the same
token, the measure of the sons' rebellion is how much liquor
they can "sneak".'

Raleigh wrote that 'sneaking a drink' had more significance
for the Irish than for other cultural groups. 'Allied to this
peculiar Irish custom are the concomitant phrase and action:
watering the whiskey; that is, filling the bottle with water to the
level where it was before you sneaked your drink.' In some Irish
households, Raleigh said, 'whole cases of whiskey slowly
evolve into watery, brown liquid, without the bottles ever being
set forth socially, so to speak. This act—the lonely, surreptitious,
rapid gulp of whiskey—is the national rite . . .'

National or not, it was clearly an O'Neill family rite, as was
the medicinal use of alcohol, another custom traditionally Irish.

'The Irish addiction to drink is a simplifying element in their
lives,' wrote Raleigh. 'This is how all problems are met—to
reach for the bottle.' And reach the O'Neills did. 'A drop now
and then is no harm when you are in low spirits or have a bad

cold,' advises the maid in *Long Day's Journey*, and Eugene's father agreed. 'I've always found good whiskey the best of tonics,' he says in the play, calling drink the 'good man's failing'. Even O'Neill's mother found alcohol a 'healthy stimulant'.

According to Irish scholars, alcohol served utilitarian functions other than medicinal, at least in earlier times. The Irish were as Puritanical about sex as they were tolerant toward drunkenness, and in fact the two were linked—alcohol served as a sexual substitute. The teetotaller, indeed, was considered a menace, a man who prowled the streets and got girls into trouble. When a young man was unhappy in love, he was advised to 'drink it off'. Wrote one sociologist: 'Drowning one's sorrows becomes the expected means of relief, much as prayer is among women.'

These attitudes toward alcohol, held by at least some Irish families at the turn of the century, are a far cry from attitudes toward alcohol held by Southern Baptists or Jews. The former condemn any use of alcohol and the latter condemn drunkenness.

How these traditions develop in a given society is as complicated and hard to comprehend as is the pairing and unpairing of molecules in the microscopic world of genes. But obviously customs toward drinking, like 'prayer among women', do not easily lend themselves to a genetic explanation.

3. *Frustrated, unhappy, insecure, lonely people drink to feel less frustrated, unhappy, insecure, and lonely.* This, perhaps, is the most popular explanation of why people drink to excess. The assumption is that people drink to escape. They have stresses to which they respond by losing themselves in alcohol.

The fallacy in the reasoning is a simple one: most people are frustrated, unhappy, insecure, or lonely much of the time but do not become alcoholic. Thirty per cent of US adults do not drink at all. This is not a purely environmental explanation in any case. The question has to be asked: why do some individuals respond to stress more intensely than others? Response to stress itself may in part be genetically determined.

The same point can be made about people who drink apparently to relieve depression. Certain depressive illnesses them-

selves may be genetic; there is substantial evidence for this.
Moreover, most people probably do *not* drink to relieve depression. In a study of patients with depressive disorders, one-third
drank somewhat more when depressed but another third drank
somewhat less. In fact, depression may follow rather than lead
to drinking. In studies where individuals have been given large
amounts of alcohol, they develop depressive symptoms.

4. *Alcoholics are loners.* In a letter to the author, the historian
Gilman Ostrander proposed the following explanation for
alcoholism:

Alcoholism is basically a disease of individualism. It afflicts people who
from early childhood develop a strong sense of being psychologically
alone and on their own in the world. This solitary outlook prevents
them from gaining emotional release through associations with other
people, but they find they can get this emotional release by drinking. So
they become dependent on alcohol in the way other people are dependent on their social relationships with friends and relatives.

Ostrander believes his theory explains why alcoholism is more
prevalent in some ethnic groups than in others. The high alcoholism rate among the Irish and French, he says, is at least
partly traceable to the fact that Irish and French children are
brought up to be

. . . responsible for their own conduct. When they grow up and leave
the household, they are expected to be able to take care of themselves.
Individualism in this sense is highly characteristic of these groups.

Jews and Japanese, on the other hand, have a low alcoholism
rate because children are *not* expected to be independent.

Infants in these families are badly spoiled, that is to say, their whims are
indulged in by parents and older relatives, so that, from the outset, they
become emotionally dependent upon others in the family . . . It is never
possible for them to acquire the sense of separate identity, apart from
their family, that is beaten into children in, say, Ireland . . . They are
likely to remain emotionally dependent upon a part of their family in a
way that is not true in societies where the coddling of children is socially
disapproved of.

And this is why Ostrander believes most Jews and even hard-drinking Japanese do not become alcoholic.

They never had the chance to think of themselves as individuals in the Western sense of the word. They are brought up to be so dependent upon others in the family that they are unable to think of themselves as isolated individuals.

It is a hard theory to prove. There may even be some question about the low rate of alcoholism in Japan relative to other Far Eastern countries. Nevertheless, it is an interesting explanation for the differences that appear to be real between various cultures, and deserves attention from sociologists who might find ways of testing the theory scientifically.

In conclusion, it should be noted that nobody disputes the importance of cultural factors in drinking behaviour, and it seems entirely possible, though difficult to prove, that environmental factors have some bearing on whether a person becomes a problem drinker or an alcoholic. At the very least, alcohol must be available for alcoholism to occur; a person must drink before he can be a drunkard. Whether or not he drinks will certainly be influenced by his social environment. Whether he drinks excessively also may be influenced by his social environment. The puzzle remains: What specific factors—genetic, environmental, or both—produce serious alcohol problems in a rather small minority of drinkers?

SECTION FOUR

Treating alcoholism

Formula for longevity: Have a chronic disease and take care of it.

—Oliver Wendell Holmes

14 *The treaters*

Sometimes people recover from an illness without professional help. This is called spontaneous remission, 'spontaneous' meaning the remission cannot be explained. It happens in almost every illness, including alcoholism. Before treatment can be judged effective, it must be shown to be superior to no treatment.

Usually studies comparing treated and untreated groups are needed in order to show effectiveness. Some treatments are so effective, however, that such studies are not needed. Penicillin for pneumonia is an example. But it is a mistake to judge the effectiveness of a treatment by what happens to one, two, or a small group of patients. Even terminal-cancer patients sometimes recover 'spontaneously', and the history of medicine is a graveyard for treatments that were worthless but flourished because people tend to get over things anyhow.

There is no penicillin for alcoholism. Studies are needed. Some of the treatments discussed in the next chapter have never been studied; others have been studied, but not well. It may be a slight exaggeration, but only slight, to say that no study has proved beyond question that one treatment for alcoholism is superior to another treatment or to no treatment.

Nevertheless, alcoholics seek help and people try to help them. Uncertainty about the results of trea⁺ment has not and should not discourage this effort. Who is providing treatment? What treatments are available?

The providers include social workers, psychologists, psychiatrists, other physicians, and people called alcoholism counsellors, many of whom are recovered alcoholics.*

---

* The term 'recovered' alcoholic deserves comment. AA takes the view that no one recovers from alcoholism; they simply stop drinking. Whether or not this is always the case, one should be cautioned against using terms that seem synonymous with 'recovered' but have different connotations. Terms to be avoided are 'ex-alcoholic' and 'reformed' alcoholic, both having a criminalistic flavour.

Concerning physicians, two points of view are heard: they should treat more alcoholics; they should treat no alcoholics. The view depends on whether alcoholism is seen as an illness.

If it is, then it is generally agreed that physicians are less active in treating the illness than they should be. Psychiatrists are notoriously reluctant to see alcoholics. Many view alcoholics as people who don't pay their bills, miss appointments, call at all hours, and test the psychiatrist's patience by refusing to get better. Even when treating a patient with a drinking problem, some psychiatrists interpret the drinking as symptomatic of some other condition and ignore it.

Alcoholism is often ignored by other physicians as well. A survey of a large medical ward revealed that 25 per cent of the male patients had a serious drinking problem that may have contributed to their illness, but the hospital charts rarely mentioned drinking. Another study found that before some physicians suspect alcoholism, the alcoholic has to be dirty and unshaven. Until recent years many general hospitals in the USA would not admit patients with a diagnosis of alcoholism. Alcoholics often take a dim view of doctors. Since the feeling is often reciprocated, this attitude is understandable.

Naturally, those who believe alcoholism is *not* a disease do not expect doctors to help alcoholics and indeed would prefer that they not try (except of course for medical complications, such as cirrhosis and delirium tremens). Some alcoholics reject any treatment that does not come from other alcoholics. 'How can anyone help an alcoholic who has not been one?' goes the reasoning. The same principle could apply to any condition. How can anyone treat schizophrenia or diabetes who has not been schizophrenic or diabetic? Some feel they can, and do, with some success.

The real issue concerns not who treats alcoholics but who treats them best. Until social workers or recovered alcoholics can show they get better results than medical doctors, or vice versa, professional chauvinism seems ill-advised.

Trends

A large number of reports in recent years have indicated that alcoholic patients benefit as much from brief out-patient services as they do from in-patient care. For more than two decades in the USA, treatment of alcoholism was concentrated at facilities modelled after the so-called Minnesota Plan. Patients were admitted for a period of several weeks and spent from early morning to late in the evening participating in group therapy, AA meetings, individual therapy, and educational activities, as well as a good deal of walking and performing mild exercise. Sometimes patients spent time on the 'hot seat' where their faults were dissected with gusto by the other patients. The treatment philosophy was based on principles of AA, including the need for total abstinence. The treatment was expensive, costing upwards to $25 000. As long as insurance companies ('third parties') picked up the bill, the system became increasingly popular, and even small cities would have at least one alcoholism and drug in-patient facility. Then insurers became aware of the lack of evidence supporting the effectiveness of long-term hospitalization and began curtailing reimbursements, encouraging brief out-patient therapies. Until more evidence is obtained that in-patient treatment is superior to out-patient counselling, this trend will probably continue.

A similar trend away from lengthy hospitalization has occurred in the UK, where there has been increasing emphasis on community-based, multidisciplinary teams. The teams are usually based in a building away from any hospital; they include psychiatric nurses, a psychologist, a social worker, counsellors, and administrators/clerks. Some teams have a psychiatric consultant. The team members are involved in consultation, education, counselling and advisory services, and sometimes skills training. The staff size varies considerably, with as many as 50 members on some staffs and as few as two to three on others.

In addition, there are several hundred day centres in the UK, mostly in inner cities. Most provide food and shelter with welfare rights and accommodation problems. A few offer individual counselling, group psychotherapy, drama therapy, and relaxation groups. Finally, there are residential hostel projects scattered throughout the UK. Most provide 12 or fewer beds plus day care.

Residential projects involve stays ranging from three months to a year or more. They attract problem drinkers who are homeless and those who are single, unemployed, and male. The goal is to permit people to spend time in an alcohol-free environment and learn to break their dependence on drinking.

Since the 1960s National Health Insurance has supported treatment units (ATUs). Those that operate as separate units within hospitals are becoming fairly obsolete, with patient care shifted to the community. Serious withdrawal symptoms are treated in a general or psychiatric hospital ward, although, as is true in the USA, home detoxification has become increasingly common. Sometimes it is difficult to persuade hospitals to accept individuals for detoxification.

The AA has not dominated treatment in the UK to the extent that it has in the USA, but there are still about 1500 groups throughout England and Wales subscribing to the same philosophy as AA in the USA (as described in Chapter 16).

Another group called the Drinkwatchers has appeared in recent years, intended for people who have not yet developed serious problems from alcohol but want to do something about their present level of drinking before it gets serious. Sometimes people attend only one meeting, where they receive a copy of the Drinkwatcher's manual, which suggests ways of cutting down.

At the last count there were about twenty private clinics for people with alcohol problems. Most subscribe to the 'disease' model of alcoholism and have close links with AA.

This and other information about alcohol services in the UK can be found in the 1988 Survey of Treatment and Rehabilitation Services sponsored by the World Health Organization.

15 *Specific treatments*

One thing about alcohol: it works. It may destroy a man's career, ruin his marriage, turn him into a zombie unconscious in a hall-way—but it works. On short term, it works much faster than a psychiatrist or a priest or the love of a husband or a wife. Those things . . . they all take time. They must be developed . . . But alcohol is always ready to go to work at once. Ten minutes, half an hour, the little formless fears are gone or turned into harmless amusement. But they come back. Oh yes, and they bring re-inforcements.

—Charles Orson Gorham

This chapter reviews what is available to help the alcoholic: psychotherapy, behaviour therapy, cognitive therapy, and drugs. It tends to be critical, but there is no choice if you respect evidence. I have my own method for treating alcoholism, and it seems to work, at least for a time. I've described it in some detail.

Alcoholics Anonymous, some say the best treatment of all, is discussed in Chapter 16.

Psychotherapy

There are several schools of psychotherapy, but they have one thing in common: they involve two or more persons talking to each other, one of the discussants being the psychotherapist, a trained expert in the treatment being offered. It is important to distinguish psychotherapy from counselling. As a rule, psychotherapy costs more than counselling, takes longer, and is usually performed by qualified psychotherapists, psychiatrists, or psychologists trained in a particular method of psycho-therapy. The most popular method is called psychodynamic

therapy. Although its popularity appears to be waning, it will be
discussed in some detail.

Psychodynamic therapy is generally of two types—the classical
type and the non-classical or 'watered-down' type. Both ultimately derive from the doctrines of Freud and his followers. The
classicists (psychoanalysts) attribute mental illness to unconscious conflicts that originate in early childhood. Since drinking
is an oral activity, alcoholism is assumed to arise from oral conflicts. Small babies are more oral than anyone, since practically
their whole existence revolves around sucking a nipple. If the
sucking does not go well, conflicts develop which, unless resolved in psychoanalysis, result in an 'oral-dependent' personality (dependent because tiny infants are exceedingly dependent
on mothers). Alcoholics are said to have oral-dependent personalities. They not only drink too much, but if you go to AA meetings you find they also smoke a lot and drink gallons of coffee,
as well as tending to be extremely talkative when given a
chance. Alcoholics can overcome their oral dependency, but it
takes a long time. They must first have 'insight' about their
orality and, second, 'work out' their oral conflicts by receiving
therapy for extended periods. They are encouraged to believe
this, despite the widely held view among psychoanalysts that
alcoholism is exceedingly difficult to treat. Freud himself held
that addictions were hard to treat because, at bottom, they were
so pleasurable.

Another psychoanalytic explanation for alcoholism is that
alcoholics are latent homosexuals. The reasoning is that since
both homosexuality and drinking involve oral activity, both
have origins in oral conflicts. It explains, too, why male alcoholics seek out the masculine camaraderie of bar-rooms; they
can be with men while simultaneously denying their homo-
sexual tendencies by engaging in heavy drinking, an activity
associated with masculinity. Others hold that alcoholism is a
form of self-destruction (which it obviously is) and has the same
roots as depression. People have angry feelings toward others,
cannot express them, and therefore become angry at themselves. Self-hatred is subjectively experienced as depression or
leads to self-destructive acts such as alcoholism. Therapy con-

sists of helping a person recognize his unconscious drives and motives, which results in a happier and more mature person who does not drink as much.

Many therapists who are not technically psychoanalysts nevertheless use concepts derived from Freudian theory. Any therapist who uses such terms as 'ego defence mechanism' or 'acting out' views abnormal behaviour, whether he realizes it or not, as a product of psychological conflict. This, at bottom, is what psychodynamic means. Although everyone has experienced psychological conflict, the idea that mental illness arises from conflict remains speculative.

Another form of psychotherapy is based on the theories of Eric Berne and is called transactional analysis. It will be discussed briefly.

Dr Berne was a writer whose ideas had much in common with Freud's. Instead of id, ego, and superego, he substitutes the terms child, adult, and parent. These three *dramatis personae* of mental life are constantly feuding, just as members of a family quarrel, and the quarrels sometimes take the form of abnormal behaviour. All people play games, and 'sick' people play games calculated to make them losers. The alcoholic punishes other people by punishing himself and is the loser in the end. Transactional analysis, or TA, makes alcoholics aware of the games they play and encourages them to find other games that are less destructive.

Transactional analysis lends itself somewhat more to group therapy than does psychoanalysis or psychodynamic psychotherapy. It is often said that group therapy helps alcoholics more than individual therapy does, but there is no evidence for this. Group therapy does have one advantage: it takes less of the therapist's time.

To show that a particular treatment is useful, three questions must be asked: Would the patient have recovered without any treatment? Would he have done as well or better with another treatment? Is the improvement related to non-specific aspects of a treatment? 'Non-specific aspects' include, in Peter Medawar's words, the 'assurance of a regular sympathetic hearing, the feeling that somebody is taking his condition seriously, the discovery that others are in the same predicament, the comfort of learning that his condition is explicable (which does not depend on the explanation being the right one)'. These factors are common to most forms of psychological treatment and the good they do cannot be credited to any one treatment in particular.

Behaviour therapy

There are two kinds of behaviour therapy with two different ancestries. One comes from Pavlov, who conditioned dogs. The other comes from B. F. Skinner, who conditioned pigeons. Behaviour therapy, in other words, is conditioning therapy by another name.

Pavlov found that if dogs repeatedly heard a bell before eating, eventually bells alone would make them salivate. And if you shocked the dog's foot every time he heard a bell, he would soon respond to bells in the same way he did to shock—by withdrawing the foot.

For many years attempts have been made to condition alcoholics to dislike alcohol. Alcoholics are asked to taste or smell alcohol just before a pre-administered drug makes them nauseated. Repeated pairing of alcohol and nausea results in a conditioned response: after a while alcohol alone makes them nauseated. Thereafter, it is hoped, the smell or taste of alcohol will cause nausea and discourage drinking.

Instead of pairing alcohol with nausea, other therapists have associated it with pain, shocking patients just after they drink, or they have associated it with the panic experienced from not being able to breathe by giving them a drug that causes very brief respiratory paralysis. Others have trained patients to imagine unpleasant effects from drinking, hoping to set up a conditioned response without causing so much actual distress.

Does it work? Some degree of conditioning is usually established, but it is uncertain how long the conditioning lasts. The largest study that involved conditioning alcoholics was conducted some 40 years ago in Seattle, Washington. More than

4000 patients conditioned to feel nauseated when exposed to alcohol were studied ten to fifteen years after treatment. Half were abstinent, which is an impressive recovery rate compared with other treatments. The patients who did best had booster sessions—that is, they came back to the clinic after the initial treatment to repeat the conditioning procedure. Of those who had booster sessions, 90 per cent were abstinent. Based on this study, the nausea treatment for alcoholism would seem an outstanding success. Why hasn't it been universally accepted?

One reason is that the results can be attributed to factors other than conditioning. The patients in the study were a special group. Generally they were well-educated, had jobs, and were well off financially. They may not have received the treatment otherwise, since the clinic where they were treated is private and costs money. Indeed, the patients who did best, it turned out, had the most money. Studies of alcoholics have often shown that certain subject characteristics are more predictive of successful-treatment outcome than the type of treatment administered. These include job stability, living with a relative, absence of a criminal record, and living in a rural community. In the Seattle study there was no control group that did not receive conditioning therapy. It is possible that the select group of patients, many having characteristics that favour a good outcome, would have done as well without conditioning.

Furthermore, in conditioning treatments, motivation is important. Treatment is voluntary and involves acute physical discomfort, so presumably few would consent to undergo the therapy who were not strongly motivated to stop drinking. The Seattle study makes this point graphically clear. Those who came back for booster sessions did better than those who didn't, but another group did better still: those who *wanted* to come back but couldn't because they lived too far from the hospital. *All* of these people remained abstinent.

In 1991 the first study comparing aversion therapy with treatment that did not include aversion therapy was published. The results were similar to those reported above. The 'controls' were listed in a national registry—perhaps not the ideal controls. Still, the two groups were matched on such variables as income, and the study provided further evidence that aversion helps selected individuals.

Other studies of the Pavlovian type of therapy for alcoholism (including chemical, electrical, and verbal conditioning) have been less ambitious and the results have been mixed. To the extent they seem to help, success may be attributed to factors unrelated to conditioning, such as patient selection, patient co-operation, and so on. Another factor also may promote at least short-term success. It is called the Hawthorne effect and refers to the enthusiasm therapists often have for any treatment that is new. The enthusiasm may be infectious, and patients who are enthusiastic themselves about a particular treatment may do somewhat better, for a time, than those who are neutral or unenthusiastic.

The same considerations apply to therapy based on the work of B. F. Skinner, called operant conditioning. There is a large scientific literature based on this work, but the basic ideas are simple. People behave like pigeons in the sense that they do things which are rewarded and avoid doing things which are punished. This has led to a type of treatment known as token economy. Anything a patient does that is believed good for him is rewarded (often literally with tokens which are exchangeable for food, money, and other desirable things). Anything he does that is bad for him is punished (usually simply by withholding the reward). In this manner an attempt is made to 'shape' the behaviour of patients in directions that are beneficial to them, with the hope that the new behaviour—abstinence or controlled drinking—will permanently replace the less beneficial kind.

Does this actually happen? Most treatments of this kind take place in institutional settings, and whether the new behaviour brought about in the institution 'sticks' in the outside world is not known.

One final word needs to be said about conditioning therapies. Alcoholics, by the time they seek professional help, have already suffered bitterly from their drinking, but this has not deterred them from continuing to drink. Being made to vomit or having their hand shocked by a friendly therapist is incomparably less excruciating than the physical and mental anguish that alcoholics normally experience: the morning heaves, the shakes, the crushing weight of conscience. There is a sizeable delay, to be sure, between the drinking and the anguish, and for conditioning in the literal sense to occur the delay should be shorter. But the effects of heavy drinking are so punishing that one would expect some kind of deterrent effect. After all, some people get sick from a pork chop and thereafter avoid pork chops. The need to drink must be compelling indeed, given the infinitely greater misery that comes from drinking. This is what addiction really means, as was movingly described by William James (who incidentally had a brother who was alcoholic):

The craving for a drink in real dipsomaniacs (drunkards) is of a strength of which normal persons can form no conception. Were a keg of rum in one corner of a room and were a cannon constantly discharging balls between me and it, I could not refrain from passing before that cannon in order to get the rum. If a bottle of brandy stood at one hand and the pit of hell yawned at the other, and I were convinced that I should be pushed in as sure as I took one glass, I could not refrain.

James then gives two case-histories:

A few years ago a tippler was put in an almshouse. Within a few days he had devised various expedients to procure rum, but failed. At length, however, he hit upon one which was successful. He went into the woodyard of the establishment, placed one hand upon the block, and with an axe in the other, struck it off at a single blow. With the stump raised and streaming he ran into the house and cried, 'Get me some rum! my hand is off!' In the confusion and bustle of the occasion a bowl of rum was brought, into which he plunged the bleeding member of his body, then raising the bowl to his mouth, drank freely, and exultingly exclaimed, 'Now I am satisfied.' [There also was the] man who, while under treatment for inebriety, during four weeks secretly drank the alcohol from six jars containing morbid specimens. On asking him why he had committed this loathsome act, he replied: 'Sir, it is as impossible for me to control this diseased appetite as it is for me to control the pulsations of my heart.'

To control 'this diseased appetite' with a few sessions of conditioning therapy seems a little like attacking an elephant with a pea shooter. However, one thing can be said in favour of conditioning therapy: it is inexpensive, probably does no harm, and arises from a scientific tradition that emphasizes evidence more than faith.

Cognitive therapies

'Cognitive' means to learn or know. In recent years, several new 'cognitive' approaches to the treatment of alcoholism have been proposed. They are called by a variety of names but have two elements in common: (1) they are brief, and (2) involve trying to change a patient's way of thinking.

The therapists are usually psychologists. They reject the idea that alcoholism is an illness where the patient is the victim of genes, appetites, and emotions beyond his control.

The 'disease' model compares the alcoholic to a rider who may believe he can control his horse, but this is an illusion. The horse is the master. All the rider can do is get off: give up alcohol. Not so, says the cognitive therapist. The patient *can* guide the horse. He can change his drinking behaviour by changing his thoughts.

Here are some approaches to thought-control and the names they go by:

Cognitive-behaviour therapy has a general goal and a specific technique for controlling harmful drinking. The word behaviour simply refers to the way people behave—not conditioning, as described in the previous chapter. Looking for a better treatment for depression, the American psychiatrist Aaron Beck found that depressed people stopped being depressed if they stopped having depressing thoughts. Although this sounds easier said than done, he insisted that depressed patients did not have to have depressed thoughts and that he could help them to have undepressed thoughts. He developed a rather simple set of techniques for doing this, and cognitive therapy became widely popular for almost every known psychiatric condition, including alcoholism.

In the case of depression, studies show that cognitive therapy is as good as drugs and that a combination of cognitive therapy and drugs sometimes produces the best results of all. Evidence that it is effective for the addictions, including alcoholism, has been mixed, with some studies showing that it helps and others showing that it does not. In any case, cognitive therapy for alcoholism, going under several names, has been the most studied of all alcoholism treatments and has the advantage of being brief and therefore cost-effective. Long and expensive in-patient programmes for alcoholics (often following what is called the Minnesota plan) have been shown in many studies to be no more effective than brief and relatively inexpensive out-patient treatment. The latter usually consists of giving advice and sometimes performing cognitive therapy, as well as encouraging the patient to attend Alcoholics Anonymous, at least in the USA.

Cognitive therapy aimed at changing a person's behaviour (why it is called cognitive-behaviour therapy) often begins with attempts to help the patient practise the power of positive thinking. This, older readers will remember, was the title of a bestselling book by an American preacher many years ago. The Reverend Norman Vincent Peale told his readers to avoid negative thoughts simply by telling themselves to avoid them. Similarly, the therapist teaches the patient how to identify negative thoughts and suggests ways to replace them with positive thoughts. To some extent this is based on the assumption that people cannot have two opposing thoughts simultaneously. One cannot think, 'I am miserable and therefore must drink' and simultaneously think, 'I'm feeling pretty good, so why should I drink?' Instead the person should say, 'I am a unique individual. Just because I am the son of my father does not mean I should behave like my father.' The therapist points out 'errors of logic' that the patient can correct if he perceives the errors to be errors and practises thinking good thoughts about himself. The practice takes the form of self-statements, another way of saying the patient must learn to talk to himself. He should avoid talking to himself out loud—those around might misunderstand—but by suspending the use of his vocal cords a person can carry on a conversation with himself without seeming psychotic.

After several sessions with a cognitive therapist the patient has usually acquired a collection of self-statements, sometimes called 'bumper stickers'. Common ones include, 'There is only today' and 'All of us will be dead in a 100 years'. (The latter sounds depressing but actually can be reassuring if the person worries overtime about how others view him.) My favourite bumper sticker comes from William James: 'Wisdom is learning what to overlook.' This covers a whole range of irritations at home, at work, and driving in heavy traffic.

These are self-statements intended to help patients think better of themselves (and probably, by extension, think better of others). It presumably would help anybody who 'automatically' reacts negatively to everything and everybody around them, often including the therapist. Thinking, it turns out, need not be entirely automatic. The thinker need not be entirely a passive victim of his thoughts. Patients are sometimes amazed to learn how easily they can substitute constructive, healthy thoughts for gloomy thoughts. Probably some people possess the skill more than others (maybe because of genes), and, like all arts, the art of positive thinking requires effort and concentration. The experienced cognitive therapist can help even those congenital 'awfulizers' from interpreting whatever happens to them as awful.

One reason alcoholics feel hopeless about their drinking is that they painfully remember all the times they managed to abstain entirely or moderate their intake only to go back to heavy drinking. The serious form of alcohol-dependence called alcoholism is, by most definitions, a chronic condition characterized by frequent relapses. *Relapse prevention* is a name for one form of cognitive therapy. It involves identifying 'high-risk' situations associated with relapse. This may differ from drinker to drinker, but common high-risk situations include parties, celebrations, holidays, hunger, fatigue, and loneliness. When situations most likely to produce relapse are identified in a particular patient, coping skills are then imparted for dealing with the situations. This may involve avoiding parties or, on the contrary, learning new social skills so the person can feel comfortable at the party while nursing a club soda.

Some therapists believe this can be achieved by a sheer act of will. Others disagree and point out that many aspects of a person's social life may not be manipulatable in the office of a therapist and may be crucial in determining whether the person relapses. These critics view alcohol consumption as a *social behaviour* driven by *social forces*. Relapse is believed to represent a response to social pressures such as unemployment, marital conflict, social isolation, and peer influences. The changes the patient needs in order to avoid relapse do not occur entirely *within* the person. The therapist should find him a job, conduct marriage counselling, help him find a friend or hobby. This may involve increasing the drinker's job-finding skills, of getting him a drivers' licence, and taking out a newspaper subscription. This is sometimes called *community-reinforcement*. It requires a lot of time and effort on the part of the therapist, more than most therapists are able to provide even if inclined to do so.

Practitioners of relapse prevention do not always strive to keep the patient (or client, as he is often called) from drinking entirely. Moderation may be the goal, particularly for patients with mild forms of alcohol-dependence. Relapse prevention in the severely alcoholic person, of course, is more difficult to achieve, and much, most agree, depends on whether the alcoholic is 'prepared' to stop drinking. This is based on the observation that people seem to stop drinking when they are ready to do so, and if they are not ready then professional intervention is of little help.

As mentioned, cognitive therapy goes by many names: social skills training, behaviour self-control training, stress management, and others. They all involve changing a person's way of thinking and have the advantage of being cost-effective—when they are effective.

Drug therapy

In Sears Roebuck mail order catalogues at the turn of the century two pages were devoted to drug therapies for morphine addiction and alcoholism, respectively. The drug being sold for morphine addiction consisted mainly of alcohol; a good part of the drug for alcoholism consisted of tincture of opium, a relative of morphine. Whether morphine addicts became alcoholics as a result of the treatment, or vice versa, is not known, but it illustrates the long history of giving one drug that affects mood and behaviour to relieve the effects of another drug that affects mood and behaviour. Substitution therapy reached a pinnacle with the widespread, officially sanctioned, and probably useful substitution of methadone (an addicting substance like heroin) for heroin. Heroin itself was introduced at the turn of the century as a 'heroic' cure for morphine addiction and was also believed to be useful for alcoholism. It has not done much for either condition, although a drug that blocks the effects of heroin called Naltrexone has been reported recently to promote abstinence from alcohol.

Drugs are still widely prescribed to alcoholics. They mainly consist of drugs for anxiety, such as Ativan and Valium, and drugs for depression. The anti-anxiety drugs have some effects similar to those of alcohol—they calm and relax—and are useful in relieving the jitteriness that follows heavy drinking, so that they may be useful in stopping a drinking bout. Whether they stop the *resumption* of drinking—the test of a drug's true worth in treating alcoholism—is debatable, and many clinicians feel they do not.

These drugs are sometimes used in excess by alcoholics, and sometimes in combination with alcohol. This may not be as harmful as it sounds since they have a low range of toxicity and few people become addicted in the literal sense of needing increasingly larger amounts and having serious withdrawal symptoms when they stop taking them. Nevertheless, they have obviously contributed little to the management of alcoholism, and some clinicians feel strongly that they should not be given to alcoholics for extended periods.

With one exception, there is no evidence that antidepressant medications are useful in the treatment of alcoholism, although some alcoholics do become seriously depressed and antidepressants drugs may then be indicated for their depression. The exception is a new group of drugs that enhance the neurotransmitter serotonin in its effect on brain cells. Prozac is a member of this group. These drugs reduce drinking in animals *and* humans, supporting the 'serotonin hypothesis' described in Chapter 11. Lithium, a drug useful in the treatment of mania, has been given to alcoholics, and early reports indicated that some people benefited from it. Since early reports often indicate that a particular treatment is useful, only to be refuted by later reports, lithium therapy was viewed with both interest and scepticism. The scepticism was reinforced when a large multi-hospital study in the USA failed to find that lithium was useful for alcoholism. One clinical observation that lends support to the possibility that lithium *might* be useful is that people with manic-depressive disease often drink more when they are manic than when they are depressed.

From a theoretical viewpoint, it is interesting that anti-anxiety and most antidepressant drugs do not seem to deter alcoholics from using alcohol. There is ample evidence that these drugs do indeed relieve anxiety and depression, and if alcoholics drink because they feel anxious and depressed, one would assume that the drugs would substitute for alcohol more than they seem to do. This brings up the old question considered earlier in this book, namely, are addictions specific? Are people who are vulnerable to alcohol-abuse only vulnerable to alcohol-abuse? Evidence presented in Chapters 11 and 12 suggests this may be the case.

Other treatments that have been tried are LSD and large doses of vitamins, with no convincing evidence they help.

Perhaps the drug most commonly prescribed for alcoholism over the past 25 years is one that has no effect on anxiety or depression or apparently anything else *unless* combined with alcohol. This, of course, is Antabuse.

The drug makes people physically ill when they drink. When it was first used in the early fifties, Antabuse got a bad name for two reasons. First, like most highly touted treatments, it was not the panacea its enthusiastic supporters had hoped it would be. Second, some people taking the drug died after drinking. It was later learned that the drug could be given in smaller amounts and still produce an unpleasant reaction when combined with alcohol, but death was exceedingly rare.

Partly because of the bad reputation it obtained in the early years, it has perhaps been underused since then. The more dogmatic members of AA view Antabuse as somehow incompatible with the spirit of AA, and many alcoholics resist taking the drug on the grounds that it is a 'crutch'.

Antabuse has not been entirely popular with doctors for another reason. Some still believe they must give an Antabuse 'challenge' test before prescribing the drug for indefinite periods. This test consists of giving the patient Antabuse for a few days and then giving him a small amount of alcohol to demonstrate what an Antabuse reaction is like. The Antabuse challenge test is no longer considered necessary or even desirable. Patients can be told what the effects of Antabuse will be and this will have the same effect. One awkward aspect of the challenge test is that some patients have no reaction when given the alcohol, simply because people react very differently to both alcohol and Antabuse and the cautious doses of alcohol administered are too small to produce an effect.

The main problem with Antabuse, however, is not that patients drink after taking the drug but that they stop taking the drug because they 'forget' to take it or convince themselves the drug is causing side-effects, such as impotency.* A way to obviate this problem, at least temporarily, is described below.

Something that works, provided . . .

There is an approach to treating alcoholism that works every time, given one stipulation: the patient must do what the doctor

---

* A large number of side-effects have been attributed to Antabuse, including impotency, rashes, and psychotic reactions. These occur so infrequently that it is not possible to know whether they are caused by the drug. Impotency, as explained in Chapter 6, afflicts not only intoxicated alcoholics but sometimes sober alcoholics. A small percentage of alcoholics may have rashes and psychotic episodes whether they are taking Antabuse or not. Clinicians inclined toward scepticism sometimes wonder whether side-effects attributed by the patient to Antabuse may in fact be motivated by a desire to stop taking the drug and start drinking again. Therefore, the issue of side-effects from Antabuse taken *alone* remains open to question. Combined with alcohol, of course, there is no question that Antabuse produces serious effects.

says. In this case he must do only one thing: come to the office every three or four days.

Doctors cannot help patients, as a rule, who refuse to do what they say, so there is nothing unusual about the stipulation. Why every three or four days? Because the effects of Antabuse last up to five days after a person takes it. If the patient takes Antabuse in the office, in the presence of the doctor, they both *know* he will not drink for up to five days. They have bought time, a precious thing in the treatment of alcoholism.

This approach involves other things besides Antabuse, but Antabuse makes the other things possible. First it gives hope, and hope by the time the alcoholic sees a doctor is often in short supply. He feels his case is hopeless, his family feels it is hopeless, and often the doctor feels it is hopeless. With this approach the doctor can say, 'I can help you with your drinking problem' and mean it. He doesn't mean he can help him forever (forever is a long time) and it doesn't mean the patient won't still be unhappy or that he will become a new man. It merely means he will not drink as long as he comes to the office every three or four days and takes the Antabuse. Properly warned, he won't drink unless he is crazy or stupid, and if either is the case, he probably should not be given Antabuse.

On the first visit the doctor can say something like this:

Your problem, or at least your immediate problem, is that you have trouble controlling your drinking. Let *me* take charge; let me control your drinking for a time. This will be my responsibility. Come in, take the pill, and then we can deal with other things.

I want you to stop drinking for a month. [At this point the doctor makes a note in his desk calendar to remind himself when the patient will have taken Antabuse for a month.] After that we can discuss whether you want to continue taking the pill. It will be your decision.

You need to stop for a month for two reasons. First, I need to know whether there is anything wrong with you besides drinking too much. You may have another problem that I can treat, such as a depression, but I won't be able to find out until you stop drinking for at least several weeks. Alcohol itself makes people depressed and anxious, and mimics all kinds of psychiatric illnesses.

Second, I want you to stop drinking for a month to have a chance to see that life is bearable—sometimes just barely bearable—without alcohol. Millions of people don't drink and manage. You can manage too, but you haven't had a chance recently to discover this.

F. Scott Fitzgerald complained that he could never get sober long enough to tolerate sobriety, and at least this much can be achieved with the present approach.

It is important for the patient to see the doctor (or whatever professional is responsible for his care) whenever he comes for the pill. Patients as a rule want to please their doctors; this is probably why they are more punctual in keeping office appointments than doctors are in seeing them. In the beginning the patient may be coming, in part, as a kind of favour to the doctor.

The visits can be as brief or as long as time permits. The essential thing is that rapport be established, that the patient believe something is being done to help him, and that he stay on the wagon (he has no choice if he lives up to his part of the doctor–patient contract). Brief, frequent visits can accomplish these things.

The emphasis during the visits should be not on the pill but on the problems most alcoholics face when they stop drinking. The major problem is finding out what to do with all the time that has suddenly become available now that drinking can no longer fill it. Boredom is the curse of the non-drinking drinking man. For years, most of the pleasurable things in his life have been associated with drinking: food, sex, companionship, fishing, Sunday-afternoon football. Without alcohol these things lose some of their attraction. Who can enjoy French cooking without wine, tacos without beer, or business luncheons without martinis? The alcoholic is sure *he* cannot. He tends to withdraw, brood, feel sorry for himself.

The therapist may help him find substitute pleasures—hobbies, social activities not revolving around alcohol, anything that kills time and may give some satisfaction, if not anything as satisfying as a boozy glow. In time he may find these things for himself, but meanwhile life can be awfully monotonous.

Also the patient can bring up problems of living that tend to accumulate when a person has drunk a lot. People usually feel better when they talk about problems, particularly when the listener is warm and friendly and doesn't butt into the conversation by talking about his own problems. The therapist can help by listening even if he cannot solve the problems.

If he is a psychiatrist, he can also do a thorough psychiatric examination, looking for something other than drinking to diagnose and treat. Occasionally (not often) alcoholics turn out to have a depressive illness, phobias, or other psychiatric condition.

One thing the therapist can do is help the patient accept his alcoholism. This is sometimes difficult. Alcoholics have spent most of their drinking careers persuading themselves and others that they do not have a drinking problem. The habit of self-deception, set and hardened over so many years, is hard to break. William James describes this habit with his usual verve and concludes that the alcoholic's salvation begins with breaking it:

How many excuses does the drunkard find when each new temptation comes! Others are drinking and it would be churlishness to refuse; or it is but to enable him to sleep, or just to get through this job of work; or it isn't drinking, it is because he feels so cold; or it is Christmas Day; or it is a means of stimulating him to make a more powerful resolution in favour of abstinence than any he has hitherto made; or it is just this once, and once doesn't count . . . it is, in fact, anything you like except *being a drunkard*. But if . . . through thick and thin he holds to it that he is a drunkard and nothing else, he is not likely to remain one long. The effort by which he succeeds in keeping the right *name* unwaveringly present to his mind proves to be his saving moral act.

After a month of taking the pill and talking about problems, what happens then? The patient and doctor renegotiate. Almost invariably, in my experience, the patient decides to take the pill for another month. The doctor says okay, and this is the first step in a process that must occur if the patient is going to recover: acceptance of personal responsibility for control of his drinking.

Proceeding on a month-to-month basis is a variation on the AA principle that an alcoholic should take each day as it comes.

For years, alcohol has been the most important thing in the alcoholic's life, or close to it. To be told he can never drink again is about as depressing as anything he can hear. It may not even be true. Studies indicate that a small percentage of alcoholics return to 'normal' drinking for long periods. They tend to be on the low end of the continuum of severity, but not always. 'Controlled' drinking is probably a better term than 'normal' drinking, since alcoholics continue to invest alcohol with a significance that would never occur to the truly normal drinker.

Many people, especially some AA members, reject the notion that alcoholics can ever drink normally. If alcoholism is defined as a permanent inability to drink normally, then obviously any person able to drink normally for a long period was never an alcoholic in the first place. The issue is really a definitional one, and those few alcoholics who reported sustained periods of controlled drinking in the studies were at any rate considered alcoholic when they *weren't* drinking normally. Most clinicians would agree that it is a mistake to encourage a severe classical alcoholic to believe he can ever again drink normally, but on the other hand telling him he can never drink again seems unnecessary and may not be true in every case.

When does treatment end? The minimum period is one month because that is the basis for the doctor–patient contract agreed upon in advance. Ideally, however, the treatment should continue for a minimum of six months, with the patient himself making the decision to continue taking Antabuse on a month-to-month basis. Why six months? Because there is evidence that most alcoholics who begin drinking again do so within the first six months following abstention.

A general rule applies here: the longer a patient goes without drinking at all, the shorter the relapse if a relapse occurs. It takes time to adapt to a sober way of life. Both the doctor and patient should be prepared for relapses. Alcoholism, by definition, is a chronic relapsing condition, although relapses are not inevitable. It resembles manic-depressive disease in this regard and also has similarities to such chronic medical illnesses as diabetes and multiple sclerosis. When the alcoholic has a relapse, his physician often feels resentful. When his diabetic patient has a relapse because he failed to take insulin, the doctor tends to be more understanding. The reason for this inconsistency is not clear.

Three objections have been raised concerning the above approach to treating alcoholism. The treatment is said to be based on fear, namely, the fear of getting sick, and fear is held to be one of the least desirable forms of motivation. This is debatable. Fear may be the *only* reason some alcoholics stop drinking. There is evidence that internists have somewhat better success in treating alcoholics than psychiatrists do, and the reason may be that they are in a better position to frighten the patient. They have merely to examine his liver and tell him he may be dead in a year if he keeps on drinking. Innumerable alcoholics have stopped drinking because they were told something like this. Others have stopped because they were afraid of losing their wives or jobs. It is probably no coincidence that the hardest alcoholics to treat are those who have little to lose, those who have already lost their wives, jobs, and health. They have no hope of regaining these. All they have left to lose is their life, and, by now, living has little appeal. Probably the most effective alcoholism-treatment programmes are run by industries, where the patient is an employee and his job depends on staying sober. The quotation at the beginning of this chapter is from Gorham's fictional study of an alcoholic, *Carlotta McBride*. As Gorham says, alcohol works. It has worked for the alcoholic for many years. Unless he is very much afraid of *something*, he probably will not give it up.

The second objection to the approach outlined here is that the patient becomes too dependent on a personal relationship with an authority figure, the physician, which must end at some point. In the treatment of alcoholism, the goal is not so much a lifetime cure (although sometimes this happens) as it is to bring about improvement. If the patient stays sober for longer periods after treatment than he did before, the treatment has been at least a limited success. The physician in any case should discourage a dependent relationship. He can insist upon the patient taking the pill and staying dry for a month (realizing that a month is an arbitrary unit of time and any fixed interval will do), but after that the patient has to realize that he himself has the ultimate responsibility for the control of his drinking.

The issue of dependence on authority is particularly relevant for the UK where there is evidence that *supervised Antabuse* is an effective treatment, especially when the supervisor is someone close to a patient, such as a spouse or friend. Turning over supervision to a person who actually lives with the patient is a big advantage. It reduces the number of office visits and also can be carried on for much longer than six months. Sometimes a 'contract' is signed by the therapist, patient, and supervisor. The supervisor may actually watch the patient take the Antabuse, but this is not usually necessary. If the patient starts drinking again, obviously he has not been taking the pill. Standing over an adult and watch him take a pill may have an infantalizing effect on the patient and cause resentment—sometimes sufficient resentment to excuse more drinking. One useful clause in the contract (whether written or verbal) is that the wife promises never to mention her husband's previous drinking in any context as long as he continues taking the Antabuse. The victims in the family—wife, husband, children—remember all the bad things that happen when the drinking member of the family was drunk. It tends to leave a lasting scar on the relationship. Many alcoholics are willing to take Antabuse or do practically anything to stop the nagging and harping about past mis-behaviour.

Supervised Antabuse is a much neglected treatment modality in other countries than the UK. It should be tried elsewhere. It often seems to work. Now, back to my 'American plan' of office supervision.

Finally, the complaint is heard that this approach does not get at the root of the problem; it does not explain how the patient became an alcoholic. This is true but, in my opinion, no one can explain how a person becomes an alcoholic because no one knows the cause of alcoholism. Doctors sometimes blame the patient's upbringing and patients often blame everyday stresses. There is no way to validate either explanation. There is probably no harm in telling the patient that his condition remains a medical mystery. And despite the evidence presented in Chapter 11 it is still premature to say that he *inherited* his disease.

However, if it is ever shown conclusively that some forms of alcoholism are influenced by heredity, this would not make the prognosis less favourable or the treatment less helpful. Sometimes, when evidence for a genetic factor is presented, you hear the following: 'But if it is genetic, then you can't do anything about it'. It should be noted that adult onset diabetes is almost certainly a genetic disorder and there are excellent treatments for diabetes.

16 *Alcoholics Anonymous* (AA)

'. . . Drop the anvil!'
—AA joke (see page 153)

Since its creation in 1935 by two alcoholics, this organization has grown into a world-wide network of self-help services for alcoholics and their families. It continues to dominate alcoholism treatment in the USA, in hospital settings as well as church basements; it has fared less well in other countries, particularly in Russia and the Orient.

AA has many attractive features, including three common denominators of psychotherapy (and at no cost): the assurance of a regular sympathetic hearing, the feeling that somebody is taking one's condition seriously, the discovery that others are in the same predicament. Unlike most talking therapies, AA expends little effort in trying to explain why anyone is alcoholic. The term 'allergy' is sometimes used, but usually properly bracketed in quotation marks (alcoholism does not, of course, resemble conventional allergies at all).

There is an old idea that alcoholics must become religious in order to stop drinking, and it is true that Alcoholics Anonymous has certain similarities to a religion and that some of its members have been 'converted' to AA in the same way they would be to other religions. Its 'twelve steps', for example, have a definite religious flavour, emphasizing a reliance on God, the need for forgiveness, and caring for others.

Nevertheless, to the extent that it is a religion, AA is one of the least doctrinaire and authoritarian religions imaginable. Atheists can belong to AA as comfortably as believers. There is no formal doctrine and no insistence that anyone accept a particular explanation for alcoholism. AA gives drinkers something to do when they are not drinking. It offers occasions for the soul-satisfying experience of helping someone else. It provides companions who do not drink. And it provides hope for those who need it desperately—the alcoholic and his family—and instant help for the man who wants to get back on the wagon and can't quite make it.

This sounds like a wonderful package of services, and AA is often credited with helping more alcoholics than all the other alcoholism treatments combined. There is no way of knowing whether this is true, since the kind of careful studies needed to show it have not been done. However, most professionals working with alcoholics agree that certainly nothing is lost by encouraging them to attend AA meetings and possibly much can be gained.

How does an alcoholic arrive at his first AA meeting? There are several routes.

In despair, alone or at the urging of family or friends, he may call the number in the phone book, locate a meeting, and go by himself or be escorted. His doctor or clergyman may suggest it. He may be visited by an AA member in a hospital or prison, perhaps going to his first meeting in the institution.

'When he arrives, he is likely to be frightened, depressed and still sick from withdrawal after his last drink', writes Margaret Bean in her book *Alcoholics Anonymous*. It was written in 1975, but matters have not changed much since then.

If he comes on his own and speaks to no one, he will probably be left alone. If he approaches someone or asks for help, however, he will almost invariably get it.

Most people, at first, treat him with reserve, friendly interest and encouragement. The usual practice is for the veteran member to talk about his drinking problem but not push the newcomer to reveal his. Most members realize that the newcomer has lost his self-esteem, is overcome with guilt and remorse, and feels that his weakness is all too apparent. They do not expect him to reciprocate confessional stories. The initial transaction is the instillation of hope.

The new member is taken under the wing of some sympathetic person as a 'pigeon', or beginner. He is asked if he has admitted to himself that he has trouble with alcohol, and whether he has accepted his problem. 'Admitting' and 'accepting' are carefully distinguished as separate steps. He is urged to come to many meetings, preferably

'90 meetings in 90 days,' because as a newcomer he needs them. He is taught that he should listen, keep his opinions to himself, ask questions when he does not understand, and observe and imitate successful members. He often is encouraged to give up all social contacts outside AA. The express reason for this is that he is in danger from them, because they were where he 'caught' alcoholism in the first place.

While a new member is not pushed to disclose details about himself, it is suggested that it would do him good to find someone he can talk with about his drinking. The idea is that he will find it a relief to confess all the things of which he is ashamed. For some AA members, AA may totally replace non-AA social activities.

All AA groups provide members with a protected environment in which they are treated as equal, regardless of the extent of their alcohol problem. They are freed from the fear that besets new relationships in conventional society, that one's alcoholic history will be discovered and one will be rejected because of it.

Long sobriety does confer status in AA, but there are safeguards against holier-than-thou attitudes, since at some point all members were stigmatized as alcoholics and no one is in a position to point an accusing finger. . .

Recovery is divided into three stages—physical, emotional, and spiritual. Physical recovery is the first step and the one in which Twelve Steppers are most active in the AA outreach system. A call comes for help from someone who is still sobering up. A worker at Central Service will talk with him on the telephone for a while or contact a member on call for Twelve Step work and make immediate personal contact.

The Twelve Step worker usually gets a new member to a meeting within 24 hours if possible. He knows that the desire to drink is very strong in the early stages of physical recovery and encourages the newcomer to go to as many meetings as possible.

Meetings are either open—anyone who wants to come is welcome—or closed to non-alcoholics. A new member usually has a sponsor, the person in whom he chooses to confide early in his membership.

'Many AA members go to meetings with each other, see each other relapse and return, watch each other stay sober longer and longer', Dr Bean writes. 'Many apparently expect to continue to do so several nights a week for the rest of their lives. Sometimes their families sabotage them in their use of AA and sometimes support them. Most of them are deeply loyal and very grateful to AA.'

AA has two 'Bibles'. One is called *Twelve Steps and Twelve Traditions*. The other is named *Alcoholics Anonymous*, called the 'Big Book' by members. The latter is a volume of anecdotal accounts of experiences with alcohol written by early members.

The *Twelve Traditions* book describes the organization itself. It stresses that AA is not a reform movement, nor is it operated by professionals. It is financed by voluntary contributions from its members, all of whom remain anonymous. There are no dues, no paid therapists. All comers who want help are accepted as members. All groups are autonomous. AA does not endorse other enterprises or take sides in controversies. (AA has learned a lesson from the Washington Society, a self-help group of alcoholics in the mid-nineteenth century that floundered on internal dissention over antislavery and other issues.)

When Bill W., co-founder of Alcoholics Anonymous, was asked earnestly, 'How does AA work?' he was fond of answering, 'Just fine, thanks. Just fine.' Three chapters in the 'Big Book' (Chapters 5–7) explain how AA works in detail. These are devoted essentially to the 'twelve suggested steps of recovery':

1. We admitted we were powerless over alcohol—that our lives had become unmanageable.
2. Came to believe that a Power greater than ourselves could restore us to sanity.
3. Made a decision to turn our will and our lives over to the care of God as we understand Him.
4. Made a searching and fearless moral inventory of our lives.
5. Admitted to God, to ourselves, and to another human being the exact nature of our wrongs.
6. Were entirely ready to have God remove all these defects of character.
7. Humbly asked Him to remove our shortcomings.
8. Made a list of all persons we had harmed, and became willing to make amends to them all.
9. Made direct amends to such people wherever possible.
10. Continued to take personal inventory and when we were wrong promptly admitted it.

11. Sought through prayer and meditation to improve our conscious contact with God as we understood him, praying only for knowledge of His will for us and the power to carry that out.
12. Having had a spiritual awakening as the result of these steps, we tried to carry this message to alcoholics, and to practise these principles in all our affairs.

Note that only two of the twelve steps mention alcohol. More than a prescription for abstinence, the steps give a prescription for living. As one AA member stated, 'AA doesn't teach us how to handle our drinking; it teaches us to handle sobriety. Most of us knew before we came through the door of the first meeting that the way to handle our drinking was to quit. People told us so. Almost every alcoholic I know has stopped drinking at one time or another—maybe dozens of times! So it is no trick to stop drinking; the trick is how to stay stopped.'

Many AA members approach Twelve-Step 'work' with an open mind and are prepared to be flexible. 'Greater Power', for many, stands for AA itself. God may be a symbol for the mystery of the universe rather than a traditional deity.

How successful is AA?

Some years ago, the Rand Corporation in America completed the most extensive study of alcoholism ever made. Information was obtained on 85 per cent of 922 men who had received treatment in an alcoholism unit. They were followed over four years. The report found that alcoholics who *regularly* attended AA had a higher rate of long-term abstinence than all the other groups. About half were abstinent after four years. However, only 14 per cent of the patients were regularly attending AA at the end of four years. Patients had not been randomly assigned to AA groups and other forms of treatment. Conceivably, the small minority regularly attending AA after four years represented a highly motivated group that would have done as well receiving some other treatment or no treatment.

More recently, two studies have challenged the effectiveness of AA. In both studies the subjects were atypical and generalizing from the results would be unjustified. However, they should be mentioned, particularly since no other truly controlled outcome studies of AA are available. In one study, alcoholics were court-mandated to participate in AA and compared with a no-treatment control group assigned at random. No long-term differences were found. In the second study alcohol-abusing methadone-maintenance patients were assigned randomly to an AA-based therapy group, a cognitive-behavioural training group that had 'controlled' drinking as its goal, and a no-treatment group. Fewer than 20 per cent finished the study. The controlled drinking and no-treatment groups did best; the AA group showed an increase in drinking. The relevance of these studies to the overall effectiveness of AA can be seriously challenged, but, again, these are the only studies of AA conducted according to scientific principles that exist. Given its great importance in many treatment programmes, it is unfortunate that AA has not been the subject of more research.

Having said this, the fact remains that AA has received higher praise from more people than any other approach to the problem. Dr Bean continues:

In view of the fact that all speakers adhere to a formula and everyone knows how it is going to turn out, meetings are surprisingly varied and entertaining. Speakers vary in education, charm, articulateness, age, and sex. Each tells a story: How he started to drink, lost control, and began to destroy everything in his life that had been important to him. Each has his own version of hitting bottom and—despairing, disbelieving, and full of revulsion and uncertainty—coming to AA. He then describes his recovery and how he has been able to cope with his life without alcohol.

An AA member came to our medical school to talk with students. Her story gives the flavour of AA meetings around the world and passes on some good advice to young doctors-to-be:

Hi, I'm Jan—I am an alcoholic. This is the way we introduce ourselves in an AA meeting and it seems appropriate to begin that way.

To begin with the vital statistics. I have been married to a professional man for 25 years. We have three sons. I have been sober for two and one-half years. I am a registered nurse. During the period of the alcoholism I was attending the university where I earned a bachelor's degree in

sociology. After I became sober I returned to school and this semester
I will finish a master's degree in social work. I plan to work in the field
of alcoholism when I graduate.

My life before the age of 38 was wholly unremarkable. There is no
alcoholism in my immediate family, but my paternal grandfather was
clearly alcoholic. My parents had a stable marriage; I was neither
neglected nor abused as a child. I had what one could consider a normal
adolescence, I had no problems with authority, and made straight As in
school. I was socialized to be a wife and mother; my nursing career was
strictly an insurance policy, not a career. It should be clear I was raised
before the women's movement. It was a search for my own identity that
precipitated the crisis leading to the alcoholism. Those of you who have
seen the movie, 'Kramer vs. Kramer', can have some idea of how
I lapsed into dysfunction. Except that I lacked that woman's courage
to confront the problem directly and took the back door out into
depression and alcoholism.

When I became depressed I didn't understand what was happening
to me. I only knew that I was miserable and that there was something
terribly wrong with me. I went to see a psychiatrist who started giving
me antidepressants and tranquillizers. I took those drugs almost constantly until I entered an alcoholism treatment center where they took
me off cold turkey and I went through withdrawal. I saw the psychiatrist on a weekly basis except when he was on vacation or I was. At the
end of four and one-half years of traditional psychotherapy and drugs
I was still depressed, alcoholic and addicted to the drugs as well.

It is always difficult for an alcoholic to identify that moment in time
when you cross from normal to alcoholic drinking. I can see now that
I was getting into trouble when I began using alcohol as medicine. I
began to drink at bedtime to combat the insomnia that plagued me.
I drank a lot to quell the fear and anxiety that overwhelmed me. The
longer I drank, the greater the anxiety grew—a connection I was unable
to make at the time. I also drank for all the reasons everyone else
drinks—because I thought it made me feel better, any occasion seems
more festive if you drink. I drank a lot to gain confidence in myself—
something that has never been my long suit. I also used Librium for
that purpose and I can tell you that in my drugged days this would be at
least a 100 mg performance.

Let me establish my credentials by telling you that before I began my
recovery I was hospitalized twice for detoxification, I made a serious
attempt at suicide, I left my husband at one point only to have the
drinking grow even worse and I returned. I suffered innumerable black-

outs. The longest blackout I had lasted for 24 hours. I woke up on Thursday to find out it was Friday. I used to park my car in a blackout when I went to class and then couldn't find it when I came out. I hid bottles all over the house, and sometimes did that in a blackout and couldn't find them later. During the first year of my sobriety they kept turning up all over the house.

I was a binge drinker. I would have a period of weeks of sobriety or social drinking followed by several days of uncontrolled drinking. The binges grew closer and closer together. My husband learned to time them and could often tell when he could expect to return home and find me drinking again. During this time I was trying to stay sober on will power. My psychiatrist gave me Antabuse, but it didn't do any good. Usually when I found it necessary to drink again I realized I had conveniently forgotten to take the Antabuse. And then I made the dangerous discovery that I could drink within two days of stopping the Antabuse and the reaction was tolerable. Toward the end my psychiatrist tried to get me to go to AA, but I refused. Someone might see me and suspect that I was an alcoholic—and that would have been the absolute end of the world! Besides, I kept asking, what could a bunch òf ex-drunks do for me that a board-certified psychiatrist could not? The only thing my doctor told me was that it would give me someone to call in case I wanted to take a drink.

The turning point came when my husband came to the end of his rope. He sat me down for another of our many talks where we would both agree that I simply *had* to stop drinking and I would go out again and try it on will power. Only this time it was different—this time he was telling me what *he* was going to do. 'Alcohol is controlling your life and therefore it is controlling mine and *I won't have it.*' In desperation he turned to a friend of ours who is a nurse and she had found out about treatment centers and had gotten him some Al-anon literature. My doctor had never told him about that organization. My husband tried to get me to go for treatment but I did not want to go to a mental institution, so I threw myself on his mercy with pitiful tears and pleading—something that I was very good at doing. He gave in, but extracted a promise that if I took one more drink I would go to treatment. The week had not ended until I was drinking again and he carted me off to a treatment center. I tried to leave after I got there and he told me if I came home I would go back with a sheriff because he would commit me.

In forcing me into treatment my husband provided that last final ingredient for recovery from alcoholism—and that is hope. For the first time in my life I met recovering alcoholics and learned about AA.

Out of the fog of the final days of my drinking I can recall the expression on the faces of my family and psychiatrist. It was a look of infinite sadness. As though they were saying to themselves, 'She was such a nice person—too bad she's lost.' It was like being present at my own funeral. Their attitude reinforced my own feeling of hopelessness. My behaviour was perfectly understandable to them. I was an alcoholic. They seemed to know without my telling them all the helplessness, hopelessness, fear, and anger that raged inside me. And because they understood me so thoroughly I began to believe that perhaps what had worked for them could work for me as well.

There is a spiritual component to recovery from alcoholism. It is that moment of truth when the alcoholic says, 'I give up. I surrender. I will do *anything* to stop drinking and change my life.' This is commonly known as 'hitting bottom'. It is a very painful experience for the alcoholic. For now he must give up all the defenses that have been protecting him from painful reality. You feel like you're standing naked before the whole world.

Those of us who recover in AA believe that a part of this surrender is a willingness to accept belief in a Power Greater Than Ourselves, whatever one conceives that to be. For me this occurred during the second week of my recovery. I was experiencing acute withdrawal from the Librium and I was virtually paralyzed with anxiety. I was terrified that I was losing my mind and would spend the rest of my life in a mental institution. And in that, the lowest ebb of my life, I got in touch with the finititude of my being. I gave up intellectualizing about who or what God might or might not be and prayer welled up from inside me, a simple 'Help me—for I cannot help myself.' And when I reached out, help was there for me in the form of the AA program and all the alcoholics who so lovingly taught me how to put it into practice in my own life.

I was in treatment for five weeks. When I returned home I was terrified that I would slip back into my old behavior pattern and start drinking again. But I attended AA regularly and tried diligently to work the program. And one day at a time I managed to stay sober. And the days lengthened into weeks and then into months and the day came that I realized that I not only hadn't taken a drink that day—I hadn't even thought about taking a drink. The obsession was leaving me. I can't tell you what an enormous sense of relief I felt when I finally realized I don't ever *have* to drink again.

And then I came to realize that I was experiencing life differently than I ever had before. As I have come to recognize the changes in my behavior and attitudes, I have realized that without the pain of the alcoholism I would never have opened myself up to the possibility of growth. I could have muddled around in depression and self-pity for the rest of my life if the alcoholism had not forced me to deal with it. Sometimes in AA meetings you hear an alcoholic say, 'I thank God I am an alcoholic.' The first time I heard that I hadn't been sober very long and I thought the man must have been brain damaged. But I understand now what he is saying. You also sometimes hear alcoholics say, 'I had to do everything I did to be where I am today.' That is for me the epitome of self-acceptance.

I want to take a few minutes to speak to you briefly about your own attitudes toward alcoholism. The medical profession taught *me* that alcoholics are hopeless. As a student nurse I hated alcoholic patients. They tried my patience when they were admitted drunk and took up my time that I felt better spent on patients whom I considered to be really sick. I learned this attitude from the doctors and nurses who taught me. A doctor who was teaching us said, 'I want to warn you girls about the alcoholics you will be taking care of on the wards. They can charm your socks off, but don't you be fooled—you can never get an alcoholic to stay sober.' In those days I believed everything doctors told me—I was to learn better much later. But in any case I had no reason to doubt him. I never saw anyone get sober either.

And so I ask you to stay very closely in touch with your attitudes as you begin to practise medicine. If you think alcoholism is a moral problem then you won't be inclined to intervene very actively in the disease process. By a moral problem I mean if you think the alcoholic can quit drinking by himself using will power, if he will only get his act together.

I believe alcoholism is a disease in the sense that it is something that happens to you—just as cancer or heart disease happens to you. I did not ask to become alcoholic. That was never one of my goals in life. But I was forced back into contact with reality one day and realized that I had become the victim of a process with a known symptomatology, a predictable course and a terrible prognosis. That process is called alcoholism.

We don't know what causes it, but not knowing what causes a disease has never before stopped the medical profession from treating it. I am reminded of a joke you sometimes hear around AA meetings. This drunk wanders down to the edge of a lake carrying an anvil. He's going to swim across to the other side. He jumps in and very soon he's drowning. Now on the opposite shore are all these people who want to help him. The ministers are yelling, 'We're praying for you, we're praying for you'. The doctors are yelling, 'We're doing research, we're doing research'. The AA members are yelling, 'DROP THE ANVIL!'.

Would non-alcoholic problem drinkers benefit from AA? Since they rarely attend, the question is unanswerable.

17 *Attacking the problem*

Prevention of alcoholism is an impossible dream, people say, and maybe they are right. Fifty years ago prevention of polio was an impossible dream. People said it, and they were wrong. It is true that prevention of alcoholism is still at an 'iron-lung, avoid-swimming-pools' stage, but it may not always be that way.

What can any of us—the family, employer, doctor, or society—do to help prevent alcoholism or arrest it at an early stage?

What can the family do?

First, it can recognize the problem when there is one. There often is great reluctance to do so. Wives sometimes get maternal gratification from caring for drinking husbands—it makes them feel superior, like head of the family. The children may prefer a tipsy happy daddy to a sour sober father. First the problem has to be seen as a problem—but then what?

Nothing is more frustrating for someone in the helping business (doctor, social worker, etc.) than to get a call from the spouse, 'John is drinking too much. What can I do?'

'Well have him come see me.'

'But he *won't*. He doesn't think he has a drinking problem. I'm desperate. What should I do?'

Call the police? What can the police do? Usually nothing. Drinking is not a crime. Wife-beating is, and that is when the police may help (but usually not help with the drinking).

Nag? Nagging just provides another reason to drink.

Threaten? Well, yes, sometimes. If the last straw is really the last straw, it is probably a good idea to say so. Sometimes people do stop drinking because a husband or wife threatens to leave them. Coercion sometimes works. However, often it isn't the last straw, but next to the last straw, and what then? If threats or importunings don't work, what will? Family life poses few problems as painful as this one. How the spouse handles it says a good deal about the spouse, and also about the role of the sexes in dealing with each other's problems.

A sociologist, Dr Jacqueline Wiseman, sums up a lot of experience with husbands and wives in the table on the next page. There are infinite variations, of course. 'Happy families are all alike', wrote Tolstoy. 'Every unhappy family is unhappy in its own way.' In their own ways, husbands and wives play out the roles society and their own unique personalities assign to them, and usually there is not as much 'choice' as people like to believe. Advice, even from the wisest adviser, may be bad advice simply because family relations are tremendously complicated, played out intuitively by and large, and outsiders never know how it *really* is.

Advice to be tender or tough (to leave him or not) is usually best not given, and most people don't listen anyway.

Here are some general principles to remember.

1. The alcoholic must face the consequences of his behaviour.

The family often tries to protect him from these consequences and shouldn't.

Don't pick up the pieces. If he passes out, leave him there (if it's indoors, etc.). If he throws up, let him clean it up the next morning. If he doesn't remember how the window was broken, tell him later. Be matter-of-fact. Don't pretend it was funny. Don't say I told you so. (Blackouts are scary. Sometimes people stop drinking because of them. Don't let him forget he forgot.)

Don't buy him drink.

Don't call the boss to say he has the flu (a hard rule to follow when the family depends on the income).

Don't bail him out of jail—or anything else. Let *him* explain—not you. Let *him* apologize—not you.

Stop trying to control his drinking behaviour. You can't anyway. You are as powerless in this regard as he is. Stop playing games. Stop hiding bottles. Stop pouring them down drains. Stop organizing the family routine around his drinking; shortening the cocktail hour won't help. Stop babying him. Allow *him* to be responsible for *his* behaviour. Love the sinner but not the sin.

Comparison of behaviour of wives and husbands of alcoholics*

Wives of alcoholics	Husbands of alcoholics
1. Wife notices symptoms of alcoholism early. Does not go by official symptoms like morning drinking and blackouts but by antisocial behaviour, such as: • Stays out all night • Stops taking wife out • Is rough during intercourse • Starts drinking at work	1. Husband notices much later: • Usually goes by official symptoms (often claims to know them because he is a former alcoholic himself) • Sometimes had to be told by friends (wife never had to be) • Only became really upset when wife started letting down on child care, house-work, etc. • May have noticed because wife embarrassed him in front of guests, etc.
2. Wife starts immediate campaign to get husband to stop drinking. Tries: • Logical discussion • Tears and persuasion • Nagging • Threats	2. Husband does nothing. Usually knows AA credo. Believes he can do nothing to help her stop drinking—that it is up to her to want to stop
3. Wife tries very hard to get husband into professional treatment. Often threatens to leave if he does not go. He often goes because he is so sick (physically)	3. Husband does not urge wife to go into professional treatment as soon or as strongly • Worries about the cost (not covered by insurance) • Says that she doesn't like to leave home (a few suggest that she might try weekend treatment) • Is afraid that she will meet some other man and be unfaithful if the institution takes both male and female patients
4. Wife stays with husband through entire treatment: • Many for financial reasons—whether real or imagined • AA for the insurance • For love	4. Husbands leaves wife after she slips once or twice after being institutionalized; looks for another woman (often looks while still married)
5. Wife drives husband to detoxification; picks him up	5. Husband often refuses to drive wife to detoxification centre, social worker has to call and beg him to come get her

* From Jacqueline Wiseman, *Alcohol and Women*. NIAAA Research Monograph No. 1, p. 112. Rockville, Maryland (1980).

2. Don't preach. It doesn't help.

3. Keep up hope. Many alcoholics just up and recover, with help or without.

4. Save yourself.

In the USA, try the Yellow Pages. AA usually has a listing and so does Alanon. AA usually won't send someone to the house (the alcoholic must first ask) but the spouse can go to Alanon and the kids can go to Alateen. If there is a National Council on Alcoholism in town, it can direct you to these groups. A minister can usually direct you. A physician? Less often, sad to say.

The alcoholic is isolated and ashamed. The alcoholic's family is isolated and ashamed. The kids don't invite friends over because father may be drunk. Bowling and Saturday night bridge become things of the past.

Isolated, ashamed, and bewildered, the family thinks it has never happened to anyone else. It has. Alanon and Alateen are opportunities to find this out. They are opportunities to learn how others deal with the problems. They are opportunities to learn how you may be contributing to the problem yourself without knowing it.

Should the family directly confront the alcoholic—lay it down straight and simple? Confrontation has become a big word in the treatment field. Some favour it more than others. It seemed to work with Betty Ford, wife of the US president Jerry Ford. Jerry and the children descended on her one day and said she was drinking too much. They brought along the head of a nearby alcoholism hospital. He gave her a copy of the book *Alcoholics Anonymous* and suggested she substitute the word 'chemical dependence' in the book for 'alcoholism'. 'I was in shock', writes Mrs Ford in her autobiography. She cried. She was enraged. But she went in the hospital, stayed a month, and returned home free of her chemical dependence. Forever? In Mrs Ford's case, it seems quite likely.

Confrontation is tricky. It works in some families, fails in others. Alanon, Alateen, a good alcoholism counsellor, can help a family decide whether to try.

The word 'alcoholic' has been used frequently in this book. But 'problem drinkers'—drinkers whose drinking behaviour does

not conform to the stereotype of the alcoholic (see Chapter 5)—
are far more common than alcoholics. They still have problems
caused by their drinking: their work, relationships, and family
life will suffer; they may become violent, or drink and drive; and
they will probably have health problems associated with their
drinking.

Some 'problem drinkers' become very drunk only once a year,
or once a month. Others drink steadily, every evening (sometimes lunchtimes as well), but are only rarely noticeably drunk.

Alcohol has a disinhibiting effect on behaviour. Arguments
flare up more easily, the drinker may be violent, or embarrass
their spouse, companions, or children. Remonstrations to stop
drinking, or attempts at logical discussion, are pointless. The
drinker's behaviour when drunk should be the subject of serious
discussion when they are sober, including a discussion of
possible counselling or other therapy. Sometimes this will in
practice take the form of marital therapy, during which the
drinker's problems can be discussed and the spouse or partner
helped to cope.

If the father or mother abuses alcohol in either of the ways
mentioned above, what should the children be told? What
should their attitude be towards drinking?

The parents must assume that teenagers are almost always
influenced more by peers than family, and therefore the family's
influence is of limited value. However, if there is alcoholism in
the family, the children should be informed that their own
chances of becoming alcoholic are increased and they should
be careful with alcohol, just as they should be aware of their
increased tendency to diabetes if they have adult diabetes in
the family. The situation with younger children is much more
difficult as the drinker's spouse will naturally not be inclined
to disillusion the children about their own father or mother.
However, information should be shared with them in a simpli-
fied form appropriate for their age, emphasizing if possible that
most of the time they have a wonderful father/mother. Unfortunately, it will probably be necessary to warn the children not
to mention the problem to their friends, or at school. The
drinker should be made aware of the difficult position in which
he is putting his family.

Much of what the parents tell the children about drinking will be shaped by their own religious and moral attitudes and there is no categorical advice that can be given to parents in general. The fact is: most people in our society drink and do so without harm to themselves or others. There is, however, a minority of problem drinkers and alcoholics.

Violent behaviour in a drunk person is frightening to witness, especially for children, and can be very dangerous. The spouse should first and foremost protect herself and the children, by going to a neighbour or relative's house, or other refuge. Again, the behaviour should be fully discussed when the drinker is sober, and the situation explained frankly to the children as far as possible: attempts to deny or cover up the situation are not helpful.

As a general rule, the drinker should be made to be responsible for his own actions, and the spouse should not lie to the rest of the family or outsiders to protect them or save embarrassment. By the same token, do not clear up after them, do not help them to bed, do not nurse their hangover, and do not tell their colleagues they are 'ill'.

In this day and age the dangers of drunken driving are well known. A high percentage of motor accidents, and many deaths, are caused by drunk drivers. Although different blood-alcohol limits are set in different States and different countries, the simplest rule is to abstain altogether if you have to drive. The spouse or companion of a problem drinker should arrange in advance to take the car keys, or arrange a taxi home. If the drinker insists on driving against all advice, then the spouse should take a taxi home. Once again, his behaviour should be discussed later and he should be prevented by all means possible from ever drinking and driving.

Some problem drinkers will never become intoxicated (or abusive to anyone) but still drink heavily regularly, and this is likely to cause problems to their liver, stomach or pancreas, kidneys, or brain. What can be done?

If at all possible, the drinker, accompanied by his or her spouse, should see the family doctor. For the healthy but worried spouse to call the doctor without telling the offender may make the matter worse. The person may now drink as a result of resentment, among other reasons. Remember, he rarely or never gets drunk; his behaviour is normal. But he is harming his health and may practise 'denial' in the same way that alcoholics do: lie to the doctor and deceive himself.

Occasionally these people can be helped by the family doctor by following the Antabuse plan described in Chapter 16. He can tell the patient that he realizes the patient is not an alcoholic but that apparently he has a sensitivity to alcohol and should stop drinking entirely for a few weeks to see whether his heartburn, diarrhoea, or painful urination clears up. This all has to be done while preserving the patient's self-respect. In these people the spouse can usually do little more than encourage the patient to see the doctor and ask to accompany him, with the goal that the doctor has a factual account of what has been happening. In other cases a clear description by the doctor of how he is damaging his health may persuade the drinker to cut down.

A person who continues to drink despite hepatitis (inflammation of the liver) must be viewed as a compulsive drinker just as the alcoholic is a compulsive drinker. Problem drinkers are extremely resistant to suggestions that they need counselling or psychotherapy or that they have a drinking problem. For this reason the problem drinker is extremely hard to treat. The AA is right in this regard: many drinkers have to 'hit bottom', meaning they have almost ruined their lives from drinking, before they are willing to give up self-deception and look for help.

There is a very small group of people who have a condition called pathological intoxication. This interesting condition (sometimes the subject of movies) has these features: the person has two or three drinks (usually no more) and has a sudden change of behaviour, usually towards violent behaviour. A few hours later, or the next day, the person has no memory of the episode.

Pathological intoxication seems particularly prevalent among people who have a history of head injury (including former prize-fighters). It is possible, although not proven, that pathological intoxication is a manifestation of a seizure disorder, a form of 'focal epilepsy' that does not involve violent shaking of the extremities but occurs deep in the brain and reflects a neuro-logical susceptibility to smallish amounts of alcohol. In any case, if any member of the family shows this behaviour they should be encouraged to see their family doctor at once, who will probably refer them to a neurologist. Many of these people have abnormal electroencephalograms (brain waves). The only treatment is total abstinence from alcohol.

Since sufferers of pathological intoxication are usually not alcoholic, they are receptive to advice from a physician and, after learning what is wrong with them, many will not touch alcohol again.

What can the employer do?

Every year millions of pounds are lost to business and industry because of employees drinking on the job, alcohol-related absenteeism, and ineffective performance because of hang-overs. There is an old saying in America that you should never buy a car built on Mondays and Fridays, Mondays because of hangovers and Fridays because of long lost weekends.

For the employer, humanitarian and economic motives combine to provide a strong incentive for action against alcohol-ism. The employer is in a unique position to take action.

First, by definition, the drinking employee has a job and therefore something to lose. Alcoholics with nothing to lose have the worst prognosis. Second, the employer can document poor job performance; he has an objective yardstick to go by and doesn't have to rely on pleading or moralizing. He can rely on something better: the hovering axe.

For these reasons company alcoholism programmes tend to be successful. Some claim a rate of improvement as high as 85 per cent.

First comes the confrontation, 'Frank, we think you are drinking too much. How can we help?' Often there is a company physician who has gone to alcoholism courses and knows something about the problem. He may talk with Frank, put him on Antabuse, send him to AA meetings, or get him into a treatment programme. There is an excellent way to tell whether this is working: Frank gets to work on time, stops chewing mints on the job, and his performance improves.

People stop drinking because they are afraid of losing something important to them: job, wife, health, life. Losing health and life may be low on the list. Losing the good opinion of others may be first. The author John Cheever, a recovered alcoholic, said he stopped drinking not because of his liver but because of social disapproval. After years of social disapproval he couldn't take it any longer. Gradually, grudgingly, he stopped.

The same thing has happened to many others, and the employer is in a good position to make it occur. Money and good opinion go together and the employer has some control over both.

What can the doctor do?

Like the employer, the alcoholic's doctor (if there is one) is in a good position to identify a drinking problem early.

Doctors are notoriously slow to take advantage of this. Sometimes the patient has to show up drunk, jaundiced, with his liver down to his pelvis, before it occurs to the physician to ask whether he drinks.

Why so unobserving? One reason is that doctors don't know much about alcoholism. The subject isn't brought up much in medical school. Doctors don't like to see alcoholics. They don't know what to do with them when they see them. Alcoholics let themselves go; they look a mess. They don't pay their bills. Their wives call in the middle of the night. Their breath smells. They *want* to be this way, so what can you do? If it wasn't alcohol, it would be drugs or something else. They are obviously unstable people. 'Stay away from my door' is the message sent out by many doctors, and alcoholics get the message.

When asked about their drinking, they say no, I don't drink much. No more than anyone else. A couple of drinks before dinner. Maybe some wine with dinner. That's it.

And maybe that is it. For many doctors the subject seems too personal, almost too embarrassing, to bring up. Sometimes there seems to be an unspoken agreement between the patient and doctor that the subject will not be brought up or, if it is, that the whole truth is too much to expect. The patient doesn't want to talk about his drinking habits and the doctor doesn't press him.

One problem is that doctors don't know how to ask. 'You don't drink too much, do you?' is easy to concur with by saying no, and 'Do you have an alcohol problem?' prompts an equally quick denial.

How well a physician obtains a 'drinking history' depends on tact and training, but the important thing is to be non-judgmental.* 'Do you drink?' obviously is the first question. For the drinker, 'How do you drink?' is a good second question, neutral and unaccusing. Two questions that may follow are, 'Have you or your family ever been concerned about your drinking?' and 'Has drinking ever caused problems in your life?' At this stage the alcoholic may begin to lower his defences and disclose a problem.

Often the physical examination discloses (or hints at) a problem before the patient does.

---

* Even the best doctors sometimes have trouble getting a good drinking history:

Q Have you a drink problem?
A I have no difficulty in swallowing liquids.
Q I mean, have you a problem with alcohol?
A No, I can get all the booze I want.
Q I mean, do you drink more alcohol than you should?
A Why? How much should I drink?
Q What I mean is, do you drink more alcohol than is good for you?
A That depends on how much is good for me. What quantity do you recommend?
Q (Sighs) Is alcohol affecting your behaviour for the worse?
A My friends don't think so, but my wife is inclined to fuss about it.
Q So your wife has a problem with *your* drinking.
A No problem at all. She is quite clear in her mind what is wrong.
Q Thank you! I will see you again next Thursday. (Slips off his elastic-sided shoes which have begun to feel uncomfortably tight.)—B.J. Freedman.

With apologies to Jaroslav Hašek, author of *The Good Soldier Švejk, and his Fortunes in the World War*. William Heinemann, London, 1922. Translated from the Czech by Cecil Parrot, 1973. Reprinted from the *British Medical Journal*.

Arcus senilis (a ringlike opacity of the cornea) occurs commonly with age, causes no visual disturbance, and is considered an innocent condition. The ring forms from fatty material in the blood. Alcohol increases fat in the blood and more alcoholics are reported to have the ring than others their age.

A red nose (acne rosacea) suggests the owner has a weakness for alcoholic beverages. Often, however, people with red noses are teetotallers, or even rabid prohibitionists, and resent the insinuation.

Red palms (palmar erythema) are also suggestive, but not diagnostic, of alcoholism.

Cigarette burns between the index and middle fingers or on the chest, and contusions and bruises, should raise suspicions of alcoholic stupor.

Painless enlargement of the liver may suggest a larger alcohol intake than the liver can cope with. Severe, constant upper abdominal pain and tenderness radiating to the back indicates pancreatic inflammation, and alcohol sometimes is the cause.

Reduced sensation and weakness in the feet and legs may occur from excessive drinking.

Laboratory tests provide other clues. More than half of alcoholics have increased amounts of a chemical called GGT (gamma-glutamyl transpeptidase) in their blood, which is unusual in non-alcoholics. More than half have large red cells in their blood. Two-thirds have one or the other.

What can the physician *do* when he suspects a drinking problem? He can talk it over with the patient; this is often more productive than most physicians seem to believe. If the patient denies excessive drinking, the doctor can still suggest that he cut out alcohol entirely for a month or two. Perhaps he will sleep better, be less irritable, and get rid of the heartburn. In Chapter 15, there is a plan for helping someone abstain from alcohol which physicians may find useful.

Then, of course, the physician should know where to refer the alcoholic. Recent years have seen a proliferation of institutions for treating alcoholics and every family doctor should know those in his area. He also should know of out-patient services, which many studies show are as effective and less costly than in-patient services.

What can society do?

One can easily get the impression in our society that alcoholism
is funny. Drinking and drunkenness are the subject of many
cartoons in magazines and newspapers. Millions of humorous
greeting cards reflecting alcohol use are sold each year in the
USA, the second most popular topic for cards after sexual
behaviour.

The United States tried to prohibit drinking during the
twenties and early thirties, but failed. But, after 14 years, the
'noble experiment' ended for an obvious reason: people wanted
to drink.

They wanted to drink to satisfy a need, as Berton Roueché
explains:

> The basic needs of the human race, its members have long agreed, are
> food, clothing, and shelter. To that fundamental trinity most modern
> authorities would add, as equally compelling, security and love. There
> are, however, many other needs whose satisfaction, though somewhat
> less essential, can seldom be comfortably denied. One of these, and
> perhaps the most insistent, is an occasional release from the intolerable
> clutch of reality. All men throughout recorded history have known this
> tyranny of memory and mind, and all have sought and invariably
> found, some reliable means of briefly loosening its grip.*

But there is another reason why Prohibition is unpopular: it is
unfair. The rich have never been denied alcohol: Prohibition
invariably discriminates against the poor.

It also discriminates against minorities. This partly explains
the chaotic, crazy quilt pattern of drinking laws in the USA. It is
no accident that Mississippi and Oklahoma were last of the dry
states, and in most of the south restrictive and sometimes
bizarre laws still exist. These blossomed in the post-Civil Way
period out of the desire of southern whites to deny alcohol to
their former slaves.

In Oklahoma it was the Indians. Everybody knew the explos-
ive effect of firewater on Indians, who, sober, were menacing
enough.

* Berton Roueché, *Alcohol*. World Press, New York (1960).

The North also had restrictive laws. Here the foibles to defend against belonged to the newly arrived Irish, Germans, and Italians. These groups were known to be fond of alcohol and it was essential that they show up at the factory on time and with a clear head, meaning the saloons had to close early and be closed all day Sundays.

Country clubs, on the other hand, sometimes never closed.

Paradoxically, sometimes the rich and powerful encourage alcohol abuse in the lower classes. The eighteenth century gin epidemic in England came from a Government decision to sell gin for pennies to relieve a grain glut. Election day used to be the wettest day of the year for the unemployed in the USA whose vote could be bought for a drink.

Laws, of course, discriminate against the young who can't legally buy alcohol until a certain age. In at least one southern state it discriminates against women—in a sense. Men can buy alcohol at 18, but women have to be 21 *unless* they are married, when they can buy it at 16. (Girls there tend to marry young, and a married man is believed to be entitled to send his wife to fetch a bottle when he gets thirsty.)

What else can society do? From time to time support grows for measures like the following:

1. Ban advertising of alcoholic beverages.
2. Put warning labels on alcoholic beverages.
3. Increase taxes on alcoholic beverages.
4. Restrict availability of alcoholic beverages.

Some believe that alcohol advertising encourages alcohol abuse by glamourizing drinking and that it contributes to increased drinking among teenagers in particular. The beverage industry maintains that advertising at most induces people who already drink to change brands.

There is no evidence for either view. There never has been a scientific study of the effect of beverage alcohol advertising on alcohol abuse. Since alcohol abuse existed long before advertising, the mass media certainly did not create the problem. (For centuries the Russians were notorious tipplers, and advertising was banned entirely.)

What about television? A recent study focused on drinking behaviour in comedy, soap opera, drama, and police/detective programmes produced for British, American, and Canadian television. Results indicated that British television fiction had three times the amount of alcohol consumption seen in either American or Canadian programming. In spite of this more frequent portrayal of alcohol consumption, alcohol-related statistics indicated no greater level of alcohol misuse in the UK than in Canada or the USA. In fact, statistics indicated lower rates of liver cirrhosis in Britain than in Canada or the USA.

Should containers of alcoholic beverages have a warning label? The idea is not new. In 1945, the following label was proposed by a Massachusetts Legislative Special Commission:

> *Directions for use: Use moderately and not on successive days. Eat well while drinking, and if necessary, supplement food by vitamin tablets while drinking. Warning: If this beverage is indulged in consistently and immoderately, it may cause intoxication (drunkenness), later neuralgia and paralysis (neuritis) and serious mental derangement such as delirium tremens and other curable and incurable mental diseases, as well as kidney and liver damage.*

The label was rejected. Recently the United States Congress passed a law requiring more laconic labels, such as:

> *Caution: Consumption of alcoholic beverages may be hazardous to your health.*

There is no evidence they have made any difference. Alcohol consumption has decreased in most industrialized countries, whether the beverages have warning labels or not.

Can alcohol abuse be discouraged by increasing alcohol taxes? Would it help to place a large tax on distilled spirits and maintain lower taxes on beer and wine?

These tactics have been tried in various European countries, with discouraging results. The per capita consumption of alcohol has usually not decreased, although some change in beverage choice may occur at least temporarily. There is no evidence that alcoholism has declined as a result of higher taxes. Many believe that alcoholics are not deterred by the price of a beverage, and that massive price increases are unfair to light and moderate drinkers. This view was taken by a committee of the British House of Commons:

We do not recommend that alcoholic drink should, by the increase of taxation, be priced out of the reach of many more people. We consider that this step would unfairly penalize the vast majority of unaddicted drinkers, and do nothing to reduce the incidence of alcoholism amongst those who could still afford it, whilst temping those who could not towards dangerous alternative sources of intoxication.

A certain result of a massive price increase would be the expansion of a black market in alcoholic beverages, probably involving organized crime and the illicit production of alcoholic beverages. Some indigent alcoholics would turn to non-beverage alcohol, despite the dangers.

Finally, would it help to make alcohol less available? There has been a world-wide trend towards liberalization of drinking laws as shown by the increasing number of sales outlets, longer hours for sales, even mini-bars to dispense alcoholic beverages in hotel rooms. Has this resulted in increased use and abuse of alcohol? The evidence is inconclusive.

In 18 states in the United States the Government operates and owns liquor stores. States with Government-owned stores tend to have lower consumption rates than those with privately owned stores. The explanation may be that states whose voters prefer Government-owned stores are, for whatever reason, less heavy consumers of alcohol. One study concludes that consumption and alcoholism rates are more closely related to urbanism and income than to availability.

Since none of the above proposals offers much hope, what can society do to combat its alcohol problems? Over the long run, only two measures are likely to be effective: (i) A systematic, extensive, low-keyed, public education programme based on the best available knowledge about alcohol and its *real* dangers. (ii) Scientific studies directed toward explaining why some people abuse alcohol while most who drink do not. Knowing the cause (or causes) of alcoholism offers the best hope for ultimately preventing the disorder.

Self-appraisal

The most widely used screening questionnaire for detecting alcoholism is the Michigan Alcoholism Screening Test (MAST). A score of five or more puts you in the alcoholism category. Although some clearly non-alcoholic individuals will score five or above, this is unusual enough to make the test useful for screening purposes. It is published here for readers who wonder, 'Am I? Am I not?'*

1. Do you feel you are a normal drinker? (By normal we mean you drink less than or as much as most other people.) (No, 2 points)
2. Have you ever awakened in the morning after some drinking the night before and found that you could not remember a part of the evening? (Yes, 2 points)
3. Does your wife, husband, a parent, or other near relative ever worry or complain about your drinking? (Yes, 1 point)
4. Can you stop drinking without a struggle after one or two drinks? (No, 2 points)
5. Do you ever feel guilty about your drinking? (Yes, 1 point)
6. Do friends or relatives think you are a normal drinker? (No, 2 points)
7. Are you able to stop drinking when you want to? (No, 2 points)
8. Have you ever attended a meeting of Alcoholics Anonymous? (Yes, 5 points)
9. Have you ever got into physical fights when drinking? (Yes, 1 point)
10. Has drinking ever created problems between you and your wife, husband, a parent, or other near relative? (Yes, 2 points)
11. Has your wife, husband, a parent or other near relative ever gone to anyone for help about your drinking? (Yes, 2 points)
12. Have you ever lost friends or girlfriends because of your drinking? (Yes, 2 points)
13. Have you ever got into trouble at work because of your drinking? (Yes, 2 points)
14. Have you ever lost a job because of drinking? (Yes, 2 points)
15. Have you ever neglected your obligations, your family, or your work for two or more days in a row because you were drinking? (Yes, 2 points)

* From M. L. Selzer, A. Vinokur, and L. van Rooijen (1975). A self-administered short Michigan alcoholism screening test. *Journal of Studies on Alcohol* **36**, 117.

16. Do you drink before noon fairly often? (Yes, 1 point)
17. Have you ever been told you have liver trouble? Cirrhosis? (Yes, 2 points)
18. After heavy drinking have you ever had delirium tremens (DTs) or severe shaking, or heard voices or seen things that weren't really there? (Yes, 2 points)
19. Have you ever gone to anyone for help about your drinking? (Yes, 5 points)
20. Have you ever been in a hospital because of drinking? (Yes, 5 points)
21. Have you ever been a patient in a psychiatric hospital or on a psychiatric ward of a general hospital where drinking was part of the problem that resulted in hospitalization? (Yes, 2 points)
22. Have you ever been seen at a psychiatric or mental health clinic or gone to any doctor, social worker, or clergyman for help with any emotional problem, where drinking was part of the problem? (Yes, 2 points)
23. Have you ever been arrested for drunken driving, driving while intoxicated, or driving under the influence of alcoholic beverages? (Yes, 2 points)
24. Have you ever been arrested, even for a few hours, because of other drunken behaviour? (Yes, 2 points)

Perhaps as widely used as the MAST is the *CAGE*. The *CAGE* is a popular four-item screening technique for the recognition of substance abuse. It goes like this:

1. Have you ever felt you ought to *C*ut down on your drinking or drug use?
2. Have you felt *A*nnoyed about others criticizing your drinking or drug use?
3. Have you felt *G*uilty about your drinking or drug use?
4. Have you ever used alcohol or other drugs as *E*ye openers (i.e. to overcome a hangover or to get the day started)?

A positive answer to any two of these questions is suggestive of a drug or alcohol problem.

Index